Hanser **Understanding** Books
A Series of Mini-Tutorials

Series Editor: E.H. Immergut

Susan E.M. Selke

Understanding
Plastics Packaging Technology

Hanser Publishers, Munich

Hanser/Gardner Publications, Inc., Cincinnati

The Author:
Susan E.M. Selke, School of Packaging, Michigan State University,
East Lansing, MI 48823, U.S.A.

Distributed in the USA and in Canada by
Hanser/Gardner Publications, Inc.
6915 Valley Avenue, Cincinnati, OH 45244-3029, U.S.A.
Fax: (513) 527-8950
Phone: (513) 527-8977 or 1-800-950-8977
Internet: http://www.hansergardner.com

Distributed in all other countries by
Carl Hanser Verlag
Postfach 86 04 20, 81631 München, Germany
Fax: +49 (89) 98 12 64
Internet: http://www.hanser.de

The use of general descriptive names, trademarks, etc., in this publication, even if the former
are not especially identified, is not to be taken as a sign that such names, as understood by the
Trade Marks and Merchandise Marks Act, may accordingly be used freely by anyone.
While the advice and information in this book are believed to be true and accurate at the date
of going to press, neither the authors nor the editors nor the publisher can accept any legal
responsibility for any errors or omissions that may be made. The publisher makes no
warranty, express or implied, with respect to the material contained herein.

Library of Congress Cataloging-in-Publication Data
Selke, Susan E. M.
Understanding plastics packaging technology / Susan E.M. Selke.
    p.   cm. – (Hanser understanding books)
Includes bibliographical references and index.
ISBN 1-56990-234-8
1. Plastics in packaging.   I. Title.   II. Series.
TS198.3.P5S45   1997
688.8–dc21   97-36340

Die Deutsche Bibliothek - CIP-Einheitsaufnahme
Selke, Susan E.M.
Understanding plastics packaging technology / Susan E. M. Selke. -
Munich ; Vienna ; New York : Hanser ; Cincinnati : Hanser/Gardner,
1997
    (Hanser understanding books)
ISBN 3-446-18684-0

© Carl Hanser Verlag, Munich 1997
Camera-ready çopy prepared by the author.
Printed and bound in Germany by Druckhaus "Thomas Müntzer", Bad Langensalza

Introduction to the Series

In order to keep up in today's world of rapidly changing technology we need to open our eyes and ears and, most importantly, our minds to new scientific ideas and methods, new engineering approaches and manufacturing technologies and new product design and applications. As students graduate from college and either pursue academic polymer research or start their careers in the plastics industry, they are exposed to problems, materials, instruments and machines that are unfamiliar to them. Similarly, many working scientists and engineers who change jobs must quickly get up to speed in their new environment.

To satisfy the needs of these "newcomers" to various fields of polymer science and plastics engineering, we have invited a number of scientists and engineers, who are experts in their field and also good communicators, to write short, introductory books which let the reader "understand" the topic rather than to overwhelm him/her with a mass of facts and data. We have encouraged our authors to write the kind of book that can be read profitably by a beginner, such as a new company employee or a student, but also by someone familiar with the subject, who will gain new insights and a new perspective.

Over the years this series of **Understanding** books will provide a library of mini-tutorials on a variety of fundamental as well as technical subjects. Each book will serve as a rapid entry point or "short course" to a particular subject and we sincerely hope that the readers will reap immediate benefits when applying this knowledge to their research or work-related problems.

E.H. Immergut
Series Editor

Introduction to the Series

Preface

Understanding Plastics Packaging Technology is designed to serve as a guide for those wishing to learn about plastics packaging, how it is selected, how it is formed, etc. It can be used for introductory courses in plastics packaging, or by the practitioner who wishes to learn more about how to choose an appropriate plastic or an appropriate forming technique. This book can also be of value to the policy maker or environmentalist who wants to understand why there are such a wide variety of plastics and package forms used in packaging.

The first chapter provides some basic information about packaging which is needed to understand the functioning of plastics as a packaging material. The second chapter introduces the most significant packaging plastics and discusses their properties and major uses. The next three chapters describe, respectively, flexible packaging, packaging produced by thermoforming, and plastic packaging produced by molding processes. The processes are described relatively simply, with the focus on general understanding and packaging applications, rather than on process details. The reader interested in a deeper understanding of the processes is referred to a number of Hanser and other publications, including other books in this *Understanding* series. Chapter 6 is an overview of plastic cushioning materials and distribution packaging. Chapter 7 is a brief examination of printing and other decorating processes for plastic packaging. Chapter 8 addresses the barrier capability of plastics, including permeation, migration, and product compatibility. This subject is examined in more detail because it is nearly unique to plastic packaging. Packaging materials such as glass and metal are essentially perfect barriers to transport of volatile substances from the environment into the product, or out of the product into the environment, while paper is hardly a barrier at all. Plastics differ substantially in their ability to control this transport, and also may transfer materials between the product and the plastic packaging material. Therefore, in many applications this is a crucial area of knowledge for design and use of plastic packaging. Chapter 9 on environmental considerations is also addressed in more depth, because plastic packaging, in the last 12 years or so, has been the subject of considerable legislative attention as well as negative consumer reaction on environmental grounds. While the current climate in the U.S. is calmer, attention remains high in Europe and is growing in many other parts of the world, such as the Far East. I feel, therefore, that understanding of the environmental controversies surrounding use of plastic packaging is important in justifying its use.

I hope the reader will find this a valuable introduction to the vast field of plastics in packaging and will emerge not only with a greater understanding of this exciting field, but with a desire to go on and learn more about the many uses of plastics, and about packaging as a whole.

I wish to express my deep gratitude to my son, Erik Selke, for his efforts in producing many of the illustrations in this book.

Susan E. Selke
East Lansing, Michigan

Contents

1 Introduction to Plastics Packaging

1.1 Introduction

The purpose of this book is to introduce the reader to the field of plastics packaging. Plastics continue to increase in importance as a packaging material, replacing such materials as glass, metal, wood, and paper. A variety of different polymers are used to satisfy particular packaging needs, and a variety of different plastics processing methods are used to produce those packaging materials. This book is intended to provide an overview of the subject. References are provided so readers can learn more about particular areas of interest.

Before discussing plastics in packaging, it is necessary to briefly review packaging in general.

1.2 Functions of Packaging

The basic purpose of packaging is to enable the right goods to get to the right place at the right time in an acceptable condition. Of course, the users of packaging want to do this as economically as possible; the marketers want the package to attract consumers; the environmentalists want to minimize the environmental impacts of producing and discarding the package; and other parties have other jobs for the package to perform, as well.

Packaging dates back beyond recorded history. Primitive people used leaves, hollow gourds, and other containers to carry food and water from one place to another. Later, clay pots were used, and still later, glass and metal containers were developed. In all these cases, the fundamental function of the package was containment of the product, enabling it to be moved as a unit. This purpose is still the basic packaging function of today, but packaging has a number of additional requirements as well. There are numerous ways to classify these functions; this book uses the categories of protection, communication, and utility.

To protect its contents, a package provides some type of barrier between the enclosed product and its environment. The package either protects the product from potential damage caused by interaction with that environment, or protects something in the environment from potential damage from interaction with the product. For example, nuts oxidize and become rancid when exposed to air. Proper packaging can limit the exposure of the nuts to oxygen, thus delaying their oxidation. As another example, the accidental ingestion of aspirin by children, sometimes resulting in death, has been decreased dramatically by child-resistant packaging for most aspirin.

A term commonly used in conjunction with product protection is shelf-life. The shelf-life of a product is the length of time it remains in an acceptable condition for sale. The amount of protection provided by the package against environmental factors that can decrease product quality is, obviously, a significant determinant of the shelf-life.

The communication function of packaging comprises all the messages the package conveys to the potential purchaser or user of the product. In their most obvious form, these messages include the product's name, manufacturer, amount, directions for use, warnings, and other printed information. However, the package also conveys such non-print messages as color, shape, general image, and attractiveness. These messages are often very significant in providing brand and/or product recognition and in leading consumers to choose one product over another. For retail products, the package is often the most influential factor in the consumer's decision to purchase a particular product or brand. Thus, the importance of the package as a marketing tool must not be overlooked. Choosing package features that may increase the price of the package, and, as a result, raise the price of the product, but concurrently add sales, is an economically sound decision.

The utility function of packaging includes all those features that provide increased functionality for the user. Included in this category are dispensing features, handles, hang tabs, easy-opening features, and reclosure capability. A particular utility may not necessarily be added for the end user's (the consumer) convenience, but may assist an intermediate user, such as the retailer. A hang tab, for example, is usually provided on the packages so the retailer can easily mount them for display.

There are many other classifications of packaging functions, but they can generally be included in the categories just described. It should also be recognized that a single package feature may fit more than one of these categories at the same time. A recipe on a can of soup, for example, can provide both communication and utility.

1.3 Packaging as a System

When efficient packaging is designed, it must be viewed as only one part of a larger system. The amount of protection that a package must provide, for example, is determined by the nature of the product and the distribution system to which the product and package is exposed. The package must fit into the product manufacturing system, and be available in the needed locations in sufficient quantities. The ultimate disposal of the package is increasingly becoming a concern, as well. Thus, changes in packaging design often have far-ranging effects on the overall manufacturing/packaging/distribution/disposal system. Ideally, packaging professionals should be consulted at the product design and formulation stage, as sometimes costly packaging problems can be avoided by minute changes in the product.

1.4 Major Packaging Materials

The materials most often used in packaging are paper and paperboard, wood, glass, steel, aluminum, and plastic. Plastic competes with all of these other materials in some aspect of packaging. The advantages of plastics include low weight, durability, ease of forming, shape versatility, and economy.

The primary use of wood in packaging is in pallets, with additional use in crates and barrels. In recent years, plastic pallets have increased in use because of their durability. Plastics are also used for crates, boxes, drums and other bulk packaging systems.

Glass bottles and jars used to dominate the container market. Over the last three decades, glass has lost a major portion of its market share to plastic, with additional losses to aluminum. The major use of steel in packaging is in food cans. In many cases, changes in product formulation have caused a loss of market share to plastics, as opposed to a direct replacement of plastic containers for steel cans. Many sales of canned vegetables, for example, have been replaced by sales of frozen vegetables in flexible plastic packages. There has been dramatic growth in refrigerated and frozen food, much of it packaged in rigid and flexible plastic, at the expense of canned foods. Steel drums have lost market share to plastic drums and fiber drums.

In contrast to steel, aluminum has generally increased in market share. Aluminum took a considerable fraction of the glass beverage container market, and also has claimed a share of the steel canned fruit juice market. In closures for carbonated soft drinks, however, aluminum has lost out to plastic (specifically polypropylene). There is some evidence that the aluminum soft drink can market is beginning to suffer from the growth of plastic soft drink bottles in single-service sizes, as well.

Paper (including paperboard) is the most widely used packaging material. While in some cases, paper markets have been lost to plastics, often plastics and paper are combined to provide the desired package features. Nonetheless, here too we can find examples where plastics packaging is replacing paper.

Packaging is a major market for the plastics industry, accounting for about one-fourth of all plastics used in the United States. Low density polyethylene (LDPE) and high density polyethylene (HDPE) combined account for the largest fraction of plastics used in packaging by far, with polypropylene (PP), polystyrene (PS), polyethylene terephthalate (PET) and polyvinyl chloride (PVC) also used significantly (see Fig. 1.1). A large number of other polymers are used in smaller amounts for very specific purposes. The major packaging polymers are discussed in more detail in Chapter 2.

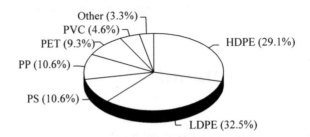

Figure 1.1 Proportion of different plastics used in packaging [1]

The largest single use of plastic packaging is in containers, including bottles, tubs, drums, pails, tubes, crates, etc. The next largest use is in film, followed by closures (caps and lids), and coatings (see Fig. 1.2).

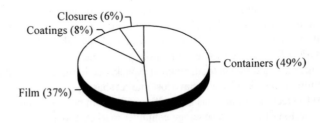

Figure 1.2 Uses of plastic in packaging [1]

1.5 Distribution Hazards

A very important packaging function is protecting goods, as they travel from the place they are produced to the end user, from deterioration in quality, breakage, etc. For many products, especially foods and pharmaceuticals, protection is achieved through packaging's barrier capabilities, which are addressed in Chapter 8. For other products, packaging must protect goods from the physical hazards associated with distribution, especially shock and vibration. Plastic foams are often used to protect products against shocks. The relationship between the product's fragility and the amount of cushioning required, as indicated by the foam's cushion curve, is discussed in Chapter 6. The fragility of a product is most often expressed as the G value (number of times the acceleration of gravity) it can sustain without failure. While fragility is specific to a given product, some general categories and their associated values are indicated in Table 1.1.

Table 1.1 Approximate Fragility of Typical Objects [2]

Classification	Example	Fragility
Rugged	Machinery	115 G and higher
Moderately rugged	Appliances	85-115 G
Moderately delicate	Televisions	60-85 G
Delicate	Electrical office equipment	40-60 G
Very delicate	Electronic equipment	25-40 G
Extremely fragile	Precision test instruments	15-25 G

A product's weight is a major factor in package design, as it affects how the product is likely to be handled. The handling method in turn influences the height from which a product may be dropped. Products which are light enough to be thrown, for example, are likely to experience significantly higher drops than products which must be handled by forklifts. Table 1.2 lists drop heights for use as guidelines in providing sufficient product protection.

A more accurate assessment of fragility takes into account the duration of the shock as well as its maximum level. The drop testing of articles at various shock levels and intensities is used to establish the product's damage boundary curve, indicating the combinations of shock intensity and duration likely to cause damage. Fig. 1.3 is a typical damage boundary curve.

Table 1.2 Maximum Expected Drop Heights [2]

Package Weight kg (lbs)	Handling Method Expected	Maximum Expected Drop Height, cm (in)
9 (20) or less	One person throws	107 (42)
9-23 (20-50)	One person carries	91 (36)
23-45 (50-100)	Two people carry	61 (24)
45-68 (100-150)	Two people carry	53 (21)
68-91 (150-200)	Two people carry	46 (18)
91-273 (200-600)	Mechanical	61 (24)
273-1364 (600-3000)	Mechanical	46 (18)
1364 (3000) and up	Mechanical	30 (12)

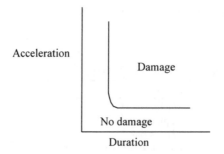

Figure 1.3 Damage boundary curve

1.6 Special Requirements of Food and Medical Packaging

Food, drugs, medical devices, and to a certain extent cosmetics need protection from contamination by microorganisms, in addition to protection from breakage and deterioration in quality. Unsterilized food products generally must be kept refrigerated to have an economical shelf-life. Products designed to be shelf-stable, requiring no refrigeration, are often processed to destroy most or all of the microorganisms they contain. A common exception is food products with low moisture contents. Moisture is necessary for the growth of microorganisms, so dried foods usually can be safely stored at room temperature.

For moist food products, the usual processing method to achieve shelf stability is retorting. Retorting typically involves packaging the product and then sending it through a steam heating process, followed by cooling. The prototype is canning. With plastic materials, this process is complicated because most plastic packaging deforms when heated, weakening seals and distorting containers. When plastic cans or pouches are substituted for metal cans or glass containers for retorted products, the process often provides a water cook step with overpressure applied to minimize the distortion of the package.

The microorganism of most concern in food products is Clostridium botulinum, which grows in anaerobic environments. It is of greater concern than more common spoilage bacteria because contamination by botulinum spores can be deadly and does not necessarily exhibit danger signs such as bad odor or taste. The spores are hardy and difficult to kill. Further, the toxin emitted is not destroyed by heat, so thorough cooking of the product after opening offers no protection.

High acid foods are less susceptible to contamination from microorganisms, including Clostridium botulinum, which does not grow in media of pH 4.6 or less. Thus, safety precautions for high acid foods such as ketchup are not as stringent as for low acid products such as peas. It is often possible to hot-fill high acid foods, rather than retort them. Plastic packaging can be used more easily in hot-fill applications.

In recent years, aseptic packaging has become increasingly common. Aseptic packaging involves sterilizing the product and package separately, and then filling and sealing the package within a sterile environment, resulting in a number of advantages. The product and package do not need to be sterilized in the same manner. Therefore the product might be sterilized with heat and the package with hydrogen peroxide. Quality can be improved because the product does not have to be heated and cooled through the package. Usually the typical geometry of a can shape means that the product near the sides is overcooked in order to ensure that the product in the center is heated sufficiently. If the product is not in the package, high temperatures, short cooking times and rapid cooling can be used effectively to maintain product quality. Plastics are used extensively in aseptic packaging.

For medical devices packaged in plastic, the two most common types of sterilization are exposure to ethylene oxide and to gamma radiation. When ethylene oxide is used, the gas is introduced into the packaged product and then evacuated after it has destroyed the microorganisms. This method of sterilization is expected to decline in use over the next few years because of the toxicity of ethylene oxide.

Packages used for ethylene oxide sterilization must be porous to gas, but not to microorganisms. In many applications, particularly in surgery, it is also important that the packages not release particulates when opened. Spunbonded polyolefins are used effectively to provide the small pores needed for the ethylene oxide movement, while preventing the entrance of microorganisms.

Radiation sterilization precludes the need for porous packages, thus permitting a wider variety of materials to be used. Both gamma and electron-beam radation are employed commercially. Other sterilization techniques include steam, gas plasma, hydrogen peroxide, periacetic acid, and X-rays, but they are not widely used for plastic packages.

Some plastics perform quite well under irradiation, but others are not suitable. Package designers should be aware that radiation-induced changes in polymer properties may take a long time to be exhibited. Exposure to radiation can start a chain of chemical events that continue long after exposure to the radiation has ceased, resulting in a loss of polymer properties.

The two basic effects of irradiation of polymers are degradation and crosslinking. These often occur simultaneously, with the overall effect on the material determined by the balance between the two. If oxygen is present during irradiation, oxidation of the polymer can significantly contribute to and accelerate degradation. Table 1.3 lists the primary effects of radiation on some important packaging polymers.

Table 1.3 Effects of Irradiation on Plastics [3]

Polymer	Primary Effect of Irradiation
Acrylonitrile-butadiene-styrene	Combination of crosslinking and degradation
Ethylene vinyl acetate	Crosslinking
Ionomers	Crosslinking
Polyamides (nylons)	Combination of crosslinking and degradation
Polycarbonate	Combination of crosslinking and degradation
Polyesters (PET)	Crosslinking
Polyethylene	Crosslinking
Polypropylene	Combination of crosslinking and degradation
Polystyrene	Crosslinking
Polyvinyl chloride	Combination of crosslinking and degradation
Polyvinylidene chloride	Degradation

Polymers which predominantly crosslink are generally suitable for use in irradiation-sterilized packaging. Properties of polyethylene, especially HDPE, are almost unaffected at typical sterilization doses. Ionomers, polyesters, and polystyrene are also little changed. High impact polystyrene is more susceptible to damage, and exhibits a decrease in impact strength, but is still usually acceptable. Irradiation of ethylene vinyl acetate tends to actually improve tensile and impact strength. In contrast, polyvinylidene chloride is not recommended for irradiation sterilization, suffering deterioration in mechanical properties and discoloration, which is especially pronounced if oxygen is present during exposure [3].

Those polymers which experience significant amounts of both crosslinking and degradation are intermediate in performance. They can generally be successfully used if they are properly stabilized and are not exposed to repeated irradiation. Again, the presence or absence of oxygen during exposure is usually an important variable [3].

1.7   Packaging Legislation and Regulation

There are a number of regulations significant to the plastics packaging industry. Environmental regulations are discussed in Chapter 9. Workplace safety and related regulations do not differ from those imposed on other industries, and are beyond the scope of this book. Additional regulations that apply to packaging include requirements for polymers used in food, drug, and cosmetic packaging; hazardous materials packaging; child-resistant packaging; and tamper-evident packaging.

1.7.1   Plastic Packaging Materials for Food, Drugs, and Cosmetics

Safety is the main reason for the special requirements for packaging in direct contact with food, drugs, or cosmetics. As is discussed in Chapter 8, components of packaging materials can migrate from the package into its contents. If the components are hazardous, the user may be harmed. In the U.S., the Food and Drug Administration (FDA) regulates most food, drugs, cosmetics, and beverages. Alcoholic beverages are also under the jurisdiction of the Bureau of Alcohol, Tobacco and Firearms (BATF), but most packaging matters are left to FDA. Meat and poultry products, including their packaging, are under the primary jurisdiction of the U.S. Department of Agriculture (USDA). However, the USDA generally defers to FDA on packaging matters, as well.

FDA regulations for pharmaceutical products are the most stringent, and those for cosmetics generally the least. There are strict rules concerning what resins may be used in specific applications, and on acceptable additives and residuals. The approval process for a new polymer to be used in food contact applications generally takes several years, because of the extensive testing required and the time needed for the FDA to act upon such petitions.. For pharmaceutical products, the process is even more complex, since the product and package are approved as a unit. Even minor package changes can require the submission of a supplementary new-drug application (NDA) for approval. Any change also requires stability testing.

For HDPE, LDPE, PET and glycol-modified PET (PETG), the United States Pharmacopoeia (USP) has formulated generic test procedures which allow pharmaceutical manufacturers to change packaging without requiring prior approval from FDA. Any HDPE resin which passes the HDPE test protocol is regarded as equivalent to the HDPE originally approved for the product, for example. Currently the fact that the use of chlorofluorocarbons (CFCs) is phasing out and consequent changes in the manufacturing process for Tyvek have led to efforts to establish a similar type of simple approval process for switching from "old" to "new" process Tyvek. Tyvek is a spun-bonded polyethylene widely used in medical device packaging. Without a simple way to handle the transition, every user would be forced, at significant expense, to seek reapproval for each type of packaged device. Similar provisions for direct material substitution are not, however, available for most other packaging materials.

While details of regulations and procedures vary considerably, most countries impose regulations to prevent hazardous packaging components from contaminating food and drugs, and to a lesser extent cosmetics.

The issue of residuals in packaging materials has been especially problematic in recent years. As our analytical capability has grown, we are able to detect minute quantities of residual substances that would have gone unnoticed a few years before. In some cases, these residuals are known to be toxic in larger amounts, but are not believed to be hazardous in the tiny quantities present in resins, and are still less dangerous in the even smaller quantities which could migrate into the package contents.

This issue came to the forefront in the U.S. in the late 1970s concerning polyacrylonitrile and PVC. In both cases, there was some residual monomer in the polymer, and the monomer was a known carcinogen. The Delaney Clause legislation, passed in 1958, prohibited the use of any known carcinogen as a food additive. Substances which migrated from packaging materials into food products were classified as indirect food additives, and were therefore subject to the Delaney Clause. On this basis, acrylonitrile copolymer beverage bottles and polyvinyl chloride liquor bottles were eliminated from the marketplace. However, the companies involved took the matter to court, arguing that the risks to consumers were negligible. In the case of the polyacrylonitrile copolymer, the residual acrylonitrile monomer levels were extremely small and the manufacturer, Monsanto, Inc., claimed there would be no migration at all [4]. In the case of PVC, modifications in technology were instituted to reduce the previously relatively high residuals to very small quantities. After several years of court battle, both cases were decided in favor of the manufacturers under what is known as the *de minimis* principle, which asserted that there are risks too small for the government to be in the business of regulating. It was calculated, for instance, that the risk of dying of cancer associated with consuming products in PVC bottles was less than the risk of being killed by a meteor.

In 1996, the U.S. government, in recognition of the changing nature of "zero," effectively replaced the Delaney Clause with a policy which examines risk. A general guideline used by the FDA is that a dietary exposure to less than 0.5 parts per billion (PPB) of a substance is below the regulatory threshold, unless there is reason to believe the substance poses a special hazard even in such small quantities [5]. This guideline is applied to potential migration of components from recycled plastics as well as to processing residuals and additives.

1.7.2 Packaging for Hazardous Materials

Packaging for hazardous materials is designed to ensure that the materials are safely contained so that they do not endanger the communities through which they travel. In addition, such packages must provide information about the contents so that emergency response personnel can appropriately handle an accident involving the release of the contents. Most nations have adopted the United Nations standards for packaging of hazardous materials based on performance standards and the classification of hazardous materials into "packing groups." Packing group I comprises materials with a high hazard level; group II covers materials with a medium hazard level; and group III deals with materials with a low hazard level. Extensive tables listing the classifications for a wide variety of hazardous materials, identification numbers, required labeling, acceptable package types, quantity limitations, and other requirements are available.

The classifications are affected by the amount of product, its type, and the type of package. For example, paint is listed as either packing group II or III, while isopentane is packing group I. For materials not included in the tables, classification is based on such criteria as flashpoint, boiling point, toxicity, and corrosion rate. All packages must pass performance tests based on the packing group and the characteristics of the material being packaged. Testing must be done by the manufacturer or in certified laboratories, and careful records kept. A system of required standardized markings on the packaging indicates the packing group, type of container, whether it is for liquids or solids, maximum allowable density or gross weight, and maximum vapor pressure for liquids. In addition, the category of material (such as flammable liquid) must be marked. The entity doing the package certification must also be indicated. There is provision for exemption from these rules for shipping small quantities of some materials. Anyone involved in the shipping of hazardous materials should seek the advice of experts in complying with the regulations.

1.7.3 Child-Resistant Packaging

In the 1960s, the U.S. Public Health Service began investigating methods to reduce the incidence of poisoning accidents among children. National Poison Prevention Week was instituted in 1961 to bring public attention to the problem, and there were a number of additional educational campaigns to get adults to keep potentially poisonous substances inaccessible to children. While some success was achieved, it was felt that there were still too many incidents occurring and a more effective approach was needed. The result was passage of the Poison Prevention Packaging Act of 1970. This legislation mandated the use of "special packaging" for products which posed a poisoning danger to young children. Such packaging helps prevent children under the age of five from obtaining access to a toxic dose of such products.

Regulations issued by the U.S. Consumer Product Safety Commission (CPSC) under the authority of this law took effect in 1973 and 1974 and required "special packaging" for most prescription drugs, many over-the-counter drugs such as aspirin and acetaminophen, and a variety of other products such as furniture polish and drain cleaner. There were provisions for exceptions for prescription drugs at the request of the purchaser, and for limited availability of other products in packages which were clearly marked "not intended for households with young children."

The U.S. Environmental Protection Agency (EPA) has equivalent requirements for some hazardous products, such as pesticides, which fall under its jurisdiction. As a result, there has been a dramatic decrease in the number of accidental ingestions and deaths associated with the covered products among young children. It was estimated that 340 children's lives were saved in the U.S. between 1974 and 1986 because of child-resistant packaging on prescription drugs alone [6]. Deaths of children resulting from accidental ingestion of aspirin fell from 46 in 1972 to none in 1985 [7].

To qualify as "special packaging," a package has to be tested with both children and adults. In the initial version of the test protocol, 90 of 100 adults tested, ages 18 through 45, must be able to successfully open the package within the test period, using only the information printed on the package. Eighty percent of 200 children tested, ages 42 through 51 months, must be unable to open the package within the 10 minute test period, even after a demonstration opening by the tester. Somewhat different requirements exist for unit dose packaging, where it is acceptable for a child to be able to open some packages provided the child is unable to gain access to a toxic dose of the product in the test period.

In the 1980s, the Consumer Product Safety Commission became concerned about the significant number of accidental ingestions which were still occurring, and studied the circumstances of these ingestions. They found that a significant number of these ingestions were from non-child-resistant packages, from products left open, or where the child-resistant feature had been disabled. This, they felt, was related to the difficulty of opening many child-resistant packages, and the inability of some adults, especially the elderly, to use these packages effectively. One result of this study was the eventual modification of the testing protocol for child-resistant packaging. Two significant changes were instituted. One provision reduced the number of children involved in the testing. If the package is effective enough with an initial group of 50 children, it is now confidently concluded that the package passed. If results are borderline, additional children can be tested. In the second provision, a test panel of older adults, ages 50 to 70, was substituted for the group of 18 to 45 year olds previously used. Many of the child-resistant packages on the market in the early 1990s could not pass when tested with older adults. A number of modifications and new package designs have been introduced as a result of this new testing protocol.

Most child-resistant packages involve plastic caps on bottles. The two most common designs include a cap that must be pressed down and turned simultaneously, and a cap with an arrow that must be lined up with one on the bottle before the cap can be snapped off. "Pull-up-and-turn" and pressing a projection inward while the cap is turned are two other designs. The first child-resistant designs frequently depended on differences in strength between children and adults. The newer, more adult-friendly designs place less emphasis on strength.

Unit-dose child-resistant packages often incorporate plastic as well. They are available in the form of blisters (formed plastic bubbles on a backing material) and pouches. The pouches most often must be opened with scissors. In blister packaging, often an individual blister must be torn off and then a small tab lifted to peel off the backing. In both cases, materials such as paper and aluminum foil are often combined with plastic for purposes of protection and strength.

1.7.4 Tamper-Evident Packaging

In 1982, seven people in Chicago died after ingesting Tylenol (Johnson & Johnson Co.) capsules containing cyanide. A malicious tamperer had inserted poisoned capsules into several bottles and returned them to store shelves, where they were bought by unsuspecting consumers. Other tampering incidents were to follow. The U.S. government then quickly acted to require "tamper-resistant" packaging for over-the-counter drug products.

The term "tamper-resistant" used in the legislation should more properly be termed "tamper-evident," since the intent is to provide the observant consumer with visible evidence that a package has been previously opened. While a number of technologies can provide this feature, the majority involve plastic packaging, including plastic overwraps around containers, plastic neck bands, and inner seals. These devices are often printed to further confound tampering.

Some researchers have raised questions about the effectiveness of various tamper-evident features. The U.S. Consumer Product Safety Commission recently called for comments on whether a performance standard should be part of the regulation, but has taken no action in this regard. Nonetheless, use of tamper-evident features has grown, particularly in the food area. There are no requirements for tamper-evident features on food packages. However, a number of manufacturers have adopted them, often positioning them as "freshness seals" or other terminology intended to convey a message to the purchaser without inciting a malicious tamperer. There is no indication that the U.S. government has any intention of extending these laws to food in the future.

Incidents of product tampering have not been confined to the United States, but other countries have not adopted similar legislation.

References

1. Modern Plastics. (Jan. 1992), pp. 53-95
2. Brandenburg, R.K. and Lee, J.J. (1985) *Fundamentals of Packaging Dynamics.* Minneapolis, MN: MTS Systems Corporation
3. Brody, A.L. and Marsh, K.S. (1997) In *The Wiley Encyclopedia of Packaging Technology, 2nd ed.*, A.L. Brody and K.S. Marsh (Eds.), New York: John Wiley & Sons, pp. 796-799
4. Lund, P.R. and McCaul, J.P. (1997) In *The Wiley Encyclopedia of Packaging Technology, 2nd ed.*, A.L. Brody and K.S. Marsh (Eds.), New York: John Wiley & Sons, pp. 669-672
5. Chemistry Review Branch, Office of Premarket Approval, Center for Food Safety and Applied Nutrition (1995) *Recommendations for Chemistry Data for Indirect Food Additive Petitions*, Washington, D.C.: U.S. Food and Drug Administration
6. U.S. Consumer Product Safety Commission (1990) *Poison Prevention Packaging: A Text for Pharmacists & Physicians*, Washington, D.C.
7. H. Lockhart (1994), Professor, School of Packaging, Michigan State University, personal communication

2   Major Packaging Polymers

2.1   Introduction

Polymers, or macromolecules, are very large molecules, formed by linkages of many smaller molecules, referred to as constitutional units, or monomers. The molecules are so large that their properties do not vary significantly with the addition or subtraction of a few of these consitutional units. Plastics, on the other hand, are materials which, at some point in their manufacture, can be shaped by heat, pressure, or both. There are many polymers, such as cellulose, which are not plastics. On the other hand, all plastics are polymers.

Plastics used in packaging, as plastics in general, can be divided into two broad groups, the thermoplastics and the thermosets. Thermoplastics are plastics which, once formed, can be melted and formed again; thermosets, once formed, do not melt and cannot be reshaped by using heat and pressure. Almost all plastics used in packaging are thermoplastics, e.g., polyethylene and polypropylene (members of the polyolefin family), polystyrene, polyvinyl chloride, polyethylene terephthalate, nylon, polycarbonate, polyvinyl acetate, ethylene vinyl alcohol, among others. Thermosets are used to some extent in coatings, particularly for cans, and to a small extent in closures. Some thermosets are also used for foam-in-place polyurethane cushioning.

Plastics can also be categorized according to the method used for their polymerization into addition and condensation polymers. Addition polymers are produced by a mechanism involving either free radicals or ions, in which small molecules are added rapidly to a growing chain, without the production of byproduct molecules. Condensation polymers are produced by the reaction of functional groups in molecules with each other in a stepwise fashion to produce a long chain polymer, and typically involve the formation of a small byproduct molecule, such as water, during each reaction step. The majority of packaging polymers, including the polyolefins, polyvinyl chloride, and polystyrene, are addition polymers.

Polyethylene terephthalate, polycarbonate, and nylon (polyamide) are some important condensation polymers.

The chemical and physical properties of plastics are influenced by their chemical composition, their average molecular weight and molecular weight distribution, their processing (and use) history, and additives. In this section, we focus on the effects of their chemical composition in determining properties and applications in packaging.

2.2   Polyethylene

About 60 % of all plastic used in packaging in the U.S. is polyethylene, largely because of its low cost, but also because of its excellent properties for many applications. High density polyethylene (HDPE) has the simplest structure of any plastic, consisting of an essentially linear moleculee with repeating ethylene units (Fig. 2.1).

$$-(CH_2CH_2)_n-$$

Figure 2.1   High density polyethylene

Low density polyethylene (LDPE) has the same chemical formula, except that it has a branched structure (Fig. 2.2).

$$-(CH_2CHR)_n-$$

where R may be -H, $-(CH_2)_mCH_3$, or a more complex structure with sub-branching

Figure 2.2   Low density polyethylene

Polyethylene, because of its very simple chemical structure, fits very easily into a crystal lattice, and hence tends to have a high degree of crystallinity. Chain branching interferes with this ability to crystallize, resulting in fewer molecules per volume, and consequently lower density

Both polymers are formed by addition polymerization of ethylene (Fig. 2.3). High density polyethylene polymerization is carried out under relatively mild reaction conditions with a

Initiation

$I\text{-}I \rightarrow I\bullet + I\bullet$
$I\bullet + M \rightarrow IM\bullet$

Propagation

$IM\bullet + M \rightarrow IMM\bullet$
$IMM\bullet + M \rightarrow IMMM\bullet$
etc.

Termination

$I(M)_n M\bullet + I(M)_p M\bullet \rightarrow I(M)_n MM(M)_p I$
$I(M)_n M\bullet + RH \rightarrow I(M)_n MH + R\bullet$
$I(M)_n M\bullet \rightarrow I(M)_n CX{=}CX_2 + H\bullet$
etc.

Figure 2.3 Addition polymerization of ethylene, where I is initiator, M is monomer, R denotes an organic species, and • represents an unpaired electron

catalyst. The catalyst does not generally produce the types of chain transfer reactions that lead to chain branching, and thus forms a predominantly linear polymer.

Two basic types of LDPE exist. The "older" type, first produced in the 1930s, is polymerized under high temperatures and pressures, conditions energetic enough to allow appreciable occurrence of chain transfer reactions which result in both long and short chain branching (Fig. 2.4). This type of LDPE is sometimes referred to as HP-LDPE (for high pressure) when it is necessary to differentiate it from linear low density polyethylene, the "younger" type of LDPE.

Short chain branching

$RCH_2CH_2CH_2CH_2\,CH_2CH\bullet \rightarrow R\overset{\bullet}{C}HCH_2CH_2CH_2\,CH_2CH_2$

$R\overset{\bullet}{C}HCH_2CH_2CH_2\,CH_2CH_2 + CH_2{=}CH_2 \rightarrow RCH(CH_2)_5$
$\qquad\qquad\qquad\qquad\qquad\qquad\qquad | $
$\qquad\qquad\qquad\qquad\qquad\qquad\quad CH_2CH_2\bullet$

Long chain branching

$RCH_2CH_2\bullet + R'CH_2CH_2R'' \rightarrow RCH_2CH_3 + R'\overset{\bullet}{C}HCH_2R''$

$R'\overset{\bullet}{C}HCH_2R'' + CH_2{=}CH_2 \rightarrow R'CHCH_2R''$
$\qquad\qquad\qquad\qquad\qquad\qquad | $
$\qquad\qquad\qquad\qquad\qquad\quad CH_2CH_2\bullet$

Figure 2.4 Short and long-chain branching reactions for polyethylene, where • is an unpaired electron

Linear low density polyethylene (LLDPE) is the newer type of LDPE. First commercially available in the mid-1970s, LLDPE is produced by polymerization of ethylene plus a comonomer under reaction conditions very similar to those used for HDPE. The comonomers result in groups on the polymer backbone that act like short chain branches, although the polymer is actually linear (see Figure 2.5). The density of LLDPE is in the same range as HP-LDPE; they are often used for the same applications, and indeed are often blended together. Therefore, it is common in the packaging literature to see references to LDPE which may mean either or both of these materials.

$$I(CH_2CH_2)_n\ CH_2CH_2\bullet + CH_2{=}CHR \rightarrow I(CH_2CH_2)_n\ CH_2CH_2\ CH_2CH\bullet$$
$$\underset{\displaystyle R}{|}$$

where R is $-CH_2CH_3$ if comonomer is 1-butene
                        $-CH_2CH_2CH_2CH_3$ if comonomer is 1-hexene
                        $-CH_2CH_2CH_2CH_2CH_2CH_3$ if comonomer is 1-octene

Figure 2.5   LLDPE polymerization

Another member of the polyethylene family is very low density polyethylene (VLDPE). This is a copolymer polyethylene with a higher comonomer content than LLDPE, giving it an even lower density, because more irregularities in the main chain further decrease the ability of the resin to crystallize.

The use of metallocene catalysts has increased the manufacturer's control over polyethylene (and polypropylene) structure, including allowing the production of "long chain branches" in otherwise linear polymers. These newest members of the polyethylene family have some unique and highly valuable properties, as discussed in section 2.2.3.

Polyethylene is generally inert, although it is attacked by oxidizing acids. Chemical resistance increases with increasing crystallinity. Polyethylene is suitable for packaging a wide variety of products. Its non-polar nature makes it a relatively good barrier against water vapor, as well as other polar materials. Since permeation occurs almost exclusively in the amorphous regions of a polymer, HDPE, with its higher crystallinity (lower percentage of amorphous regions), is a better barrier than LDPE. Neither HDPE nor LDPE are acceptable barriers for oxygen and carbon dioxide. They also do not resist the permeation of some non-polar, hydrocarbon-based compounds such as gasoline and xylene well.

Printing on polyethylene tends to be somewhat difficult because the non-polar nature of the surface does not offer binding sites for ink adherence. Therefore, if polyethylene is to be printed, its surface must usually be treated. Common surface treatments are flame treatment for containers and corona discharge treatment for films. Both of these oxidize the polymer, producing polar surface groups.

At room temperature, polyethylene is a fairly soft and flexible material. It maintains this flexibility well under cold conditions, so is applicable in frozen food packaging. However, at moderately elevated temperatures, e.g., 100 °C (212 °F), it becomes too soft for many uses. HDPE has higher brittleness and softening temperatures than LDPE, but still is not suitable for hot-fill containers.

Polyethylene meets the requirements of the FDA for food contact applications, provided any additives have been approved for such use. Polyethylene is also widely used in pharmaceutical packaging applications.

Properties of HDPE and LDPE are summarized in Table 2.1. T_g, the glass transition temperature, is the temperature above which a polymer tends to be soft and flexible, as the molecules possess segmental mobility (the ability of sections of the molecule to change their position with respect to their neighbors). T_m, the melt temperature, is the temperature at which the crystalline domains of the polymer become disordered, resulting in liquid flow. The yield represents the amount of film of a designated thickness which can be produced from a unit mass of polymer. The tensile strength is the maximum amount of stress the polymer can withstand without failure. The tensile modulus is the ratio of applied stress to strain produced in the initial linear stages of polymer deformation. WVTR is the water vapor transmission rate, measured at standard conditions.

Table 2.1 Typical Properties of Polyethylene [1-6]

Property	HDPE	LDPE
T_g, °C (°F)	-120 (-184)	-120 (-184)
T_m, °C (°F)	128-138 (262-280)	105-115 (221-239)
Density, g/cm³ (lb/ft³)	0.94-0.965 (58.7-60.2)	.912-0.925 (56.9-57.7)
Typical yield, m²/kg, 25 μm film (in²/lb, 1 mil)	41.2 (29,000)	42.6 (30,000)
Tensile strength, MPa (psi)	17.3-44.8 (2500-6500)	8.2-31.4 (1200-4550)
Tensile modulus, MPa (10³ psi)	620-1089 (89.9-158)	172-517 (24.9-75)
Elongation at break, %	10-1200	100-965
Tear strength, g/25 μm (lb/mil), film	20-60 (0.04-0.13)	200-300 (0.4-0.7)
WVTR, g μm/m² day at 37.8 °C, 90% RH (g mil/100 in² 24 h at 100 °F, 90% RH)	125 (0.32)	375-500 (0.95-1.3)
O_2 permeability, 25 °C, 10^4 cm³μm/m² day atm (77 °F, cm³ mil/100 in² 24 h atm)	4.0-7.3 (100-185)	16.3-21.3 (400-540)
CO_2 permeability, 25 °C, 10^4 cm³μm/m² day atm (77 °F, cm³ mil/100 in² 24 h atm)	20-25 (500-640)	75-106 (1900-2700)
Water absorption, %, .32 cm (125 mil) thick, 24 h	<0.01	<0.01

2.2.1 High Density Polyethylene [1, 2, 4-6]

About 30% of all plastic used in packaging in the U.S. is HDPE. It is the most widely used plastic for bottles, because of its low cost, ease of forming, and excellent performance for many applications. In its natural form, HDPE has a milky, translucent appearance, and thus, is not suitable for applications where excellent clarity is desired. However, HDPE can be readily pigmented to produce a variety of attractive colors. Common uses of HDPE include milk bottles, laundry detergent bottles, and bottles for a variety of other household chemicals, health, and beauty aid products. Other packaging applications for HDPE include 55 gallon drums, pallets, industrial containers, closures, and bags and sacks.

One drawback to using of HDPE in some of these applications is its tendency to undergo environmental stress cracking, defined as the failure of a plastic container under conditions of both stress and exposure to a product, where neither alone causes failure. Environmental stress cracking in polyethylene is strongly related to the crystallinity of the polymer. Therefore, for applications such as laundry detergent bottles, a copolymer HDPE typically containing less than 2% butene, hexene, or octene is used rather than the homopolymer. The presence of the comonomer reduces crystallinity enough to significantly improve the environmental stress crack resistance (ESCR) of the container. High molecular weight also improves resistance to environmental stress cracking.

2.2.2 Low Density Polyethylene [1, 3-5]

LDPE is the most widely used packaging polymer, accounting for about one-third of all packaging plastics in the U.S. Because of its lower crystallinity, it is a softer, more flexible material than HDPE. This property makes LDPE too soft for most bottle applications, but the material of choice for films and bags because of its low cost. LDPE has better transparency than HDPE, but still does not have the clarity desired for some packaging applications.

There are significant differences in properties between HP-LDPE and LLDPE. The long chain branching of HP-LDPE makes it a very good heat-seal material. LLDPE has a higher heat seal temperature (because of its higher melting temperature) and a narrower heat seal range. On the other hand, LLDPE has better toughness, stiffness, elongation, and barrier properties than HP-LDPE. While it generally costs more, the improved performance of LLDPE yields strong films at thinner gauges, thus producing more packages from the same amount of material. The resultant cost savings have led to the widespread replacement of HP-LDPE by LLDPE in many packaging applications. Another common approach is to use blends of HP-LDPE and LLDPE to acquire the heat-seal capability of HP-LDPE along with the performance of LLDPE. An additional advantage is improved processability, since LLDPE is usually more viscous at extrusion conditions than HP-LDPE.

The properties of LLDPE are influenced by the type as well as the amount of the comonomer used in production. Generally, performance is better, and the cost higher, in octene-based LLDPEs over hexene-based LLDPEs and hexene-based LLDPEs over butene-based LLDPEs.

The ESCR of LDPE is generally higher than that of HDPE, and LLDPE has better ESCR than HP-LDPE.

2.2.3 Metallocene Catalyzed Polyethylenes

In the last several years, a new family of catalysts for the polymerization of polyolefins has become available. Metallocene, single-site catalysts differ from traditional catalysts because they have only a single geometry for the active catalytic sites, rendering all sites equally reactive. This has profound implications for the control of molecular weight distribution and copolymer composition in LLDPE and other polyolefins.

Traditional Ziegler-Natta catalysts produce LLDPE with a relatively broad molecular weight distribution. Further, the distribution of comonomer is uneven, with the lower molecular weight fractions of the polymer containing more comonomer than average, and the higher molecular weight fractions containing less. The single-site metallocene catalysts eliminate both the high molecular weight and low molecular weight fractions, producing a polymer with a very narrow molecular weight distribution and an even distribution of comonomer. The elimination of the low molecular weight fraction results in a reduction in hexane extractables in the end product, which decreases film blocking and increases compliance with FDA requirements. The elimination of the high molecular weight fraction results in improved film clarity. Further, the polymer melt temperature becomes a much stronger function of comonomer concentration, resulting in an improved ability to tailor the properties of the end product. Such tailoring has led to decreases in the seal initiation temperature for LLDPE, allowing it to compete successfully with HP-LDPE and ethylene vinyl acetate in heat-seal applications [7, 8]. The better overall performance characteristics of the metallocene resins also can allow significant downgauging of films. However, the downside is greater difficulty in processing, because of increased viscosity and slightly lower melt strength.

Metallocene catalysts are based on a metal atom such as titanium, zirconium or hafnium, with carbon-based ligands, generally cyclopentadienyl groups, attached. Largely because of the complexity of manufacturing the catalysts, the costs of metallocene-based resins are significantly higher than those of traditional LLDPE [9], although prices are becoming more competitive. Manufacturers of metallocene-based resins include Dow Corp., Exxon Corp., Mobil Corp., and Union Carbide Corp.

A particularly significant market for metallocene resins is replacing for styrene-butadiene copolymers and ethylene-vinyl acetate copolymers, where the metallocenes can be produced at lower cost. These resins also are expected to find uses in hot melt adhesives and in some flexible PVC markets [10]. Metallocene resins are available with densities as low as 0.864 g/cm³ (54 lb/ft³), containing 20 to 40% comonomer [11].

One unique feature of metallocene catalysts is that they can produce LLDPE with long chain branches by incorporating higher alpha olefins (exceeding a hundred carbon atoms in length) into an ethylene-octene LLDPE. Traditional Ziegler-Natta catalysts are not effective with these monomers. Thus, metallocene catalyst systems can be used to produce a wide range of terpolymers containing both large and small monomers. One significant advantage is improvement in processability [12].

The resistance of polyethylene to oxygen transmission has been increased through manufacturing control of comonomer content and molecular weight distribution. Such capabilities may extend the application of polyethylene, especially for fresh produce packaging. Exxon claims their metallocene resins can extend the shelf-life of prepared salads to 21 days from the normal three days, for example [13].

While the first applications of metallocene catalysts were for LLDPE production, they have also been used in HDPE, PP, and PS manufacture. In addition to incorporating linear alpha-olefins, they can be used to add cycloolefins such as cyclobutene or cyclopentene, producing very rigid polymers with high melting points, which may compete with polycarbonates. Ethylene-styrene copolymers have also been produced with varying properties, depending on the relative amounts of the comonomers used [14-16].

2.3 Polypropylene [1, 4, 5, 17-19]

Polypropylene (PP), another member of the polyolefin family, accounts for roughly 10% of all packaging plastics in the U.S. It is an addition polymer of propylene (Fig. 2.6)

$$-(CH_2CH)_n-$$
$$|$$
$$CH_3$$

Figure 2.6 Polypropylene

PP comes in three varieties, distinguished by the spatial orientation of the methyl group on the fully extended polymer chain (see Fig. 2.7). When the orientation of the methyl group is random, the product is atactic PP. This material cannot crystallize, and as an amorphous plastic with low levels of intermolecular attraction, it has very limited use in packaging. Isotactic PP has all the methyl groups on the same side of the fully extended chain. This allows the material

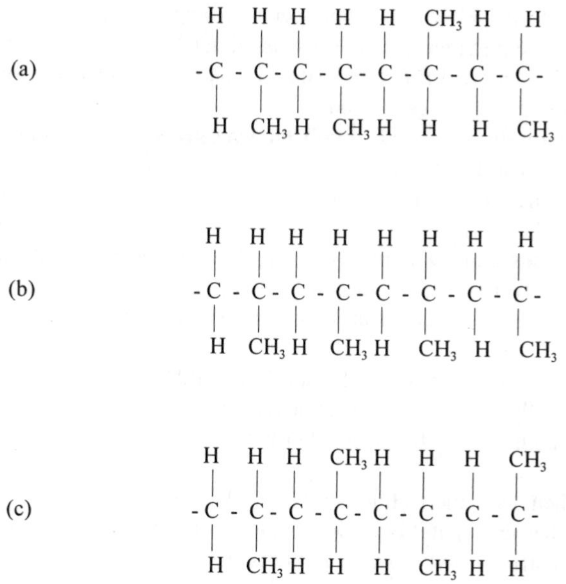

Figure 2.7 Types of polypropylene: (a) atactic, (b) isotactic, (c) syndiotactic

to fit into a crystal lattice, and the resultant crystallinity gives the material useful properties for packaging, as well as other applications. The third type has an alternating arrangement, with the methyl group first on one side and then the other. This is known as syndiotactic PP, which currently has no significant use in packaging.

PP is a lower density material than polyethylene (except for very low density polyethylene) because the presence of the methyl group causes the crystal structure to be looser. It is also a stiffer material than polyethylene, since the methyl groups interfere with ease of rotation of the main polymer chain. This increased stiffness leads to PP's widespread use in caps and closures. In such applications, polyethylene's greater softness causes it to undergo unacceptable amounts of deformation in many applications, resulting in failure of the closure to properly maintain its seal. PP's greater stiffness means it is less susceptible to deformation under load. PP also has an advantage over more rigid materials in such applications, since its flexibility allows slight undercuts to be molded in to to provide good seals. PP can also be used in linerless closures, in which resilient features within the closure provide the sealing force.

PP has a higher melting temperature than HDPE, and thus is used under conditions where HDPE becomes too soft. Typical examples include microwaveable packaging, and hot-filled bottles. On the other hand, at low temperatures, PP becomes brittle more quickly than polyethylene. Packages for frozen foods, for example, if made from homopolymer polypropylene, have a tendency to crack under impact. Incorporating a small amount of ethylene comonomer into the PP yields a polymer with improved impact strength at low temperatures. Orienting the polymer, which is discussed in Chapter 3, also improves its low temperature behavior. PP is more difficult to heat seal than polyethylene and for this reason, polypropylene films are often heat-seal coated.

PP's reduced crystallinity compared to polyethylene gives it better transparency than LDPE. The transparency can be further enhanced by adding nucleating agents to provide multiple sites for crystallite growth, thus keeping the average crystallite size very small, and by copolymerizing with ethylene to introduce structural irregularities that decreas⁻ crystallinity. These methods, along with rapid cooling of the film during production, result i.. excellent transparency. Polypropylene bottles are generally not as clear as the film, but some of the new PP copolymer resins with nucleating agents are reported to yield end products with excellent transparency.

Polypropylene was commercialized in the late 1950s. PP film's good transparency and stiffness led it to eventually replace cellophane in many packaging markets since it was less expensive, had better stability, and offered equal or better performance. Film continues to be a major market for PP.

Polypropylene is chemically inert, although it is attacked by oxidizing agents more readily than polyethylene. Its barrier capability is similar to polyethylene; it is a good water vapor barrier, but a poor gas barrier. Polypropylene has outstanding ESCR.

A useful attribute of polypropylene is its ability to undergo repeated flexing without failure. This property has led it to be called the "living hinge" material, a particularly important property for flip-top types of dispensing closures. Properties of polypropylene are summarized in Table 2.2.

The use of metallocene, single-site catalysts has produced new isotactic polypropylene resins with improved stereoregularity and controlled comonomer content resulting in higher stiffness, clarity, and melt strength. Comonomers employed to produce these enhanced PP resins include ethylene, butene, and hexene [14].

Metallocene catalyst systems may also produce syndiotactic PP with packaging applications. Ziegler-Natta catalysts cannot yield pure syndiotactic PP; generally the mix is about 95% isotactic PP, with the remainder mostly atactic PP and a very small amount of syndiotactic PP. The syndiotactic PP produced by metallocene catalysts yields a much softer, tougher, and clearer film than isotactic PP. While not expected to compete with isotactic PP, syndiotactic PP may find significant specialty markets, such as medical applications where its stability to gamma radiation is an asset [15].

2.4 Polystyrene [1, 4, 5, 20]

Polystyrene (PS) is an addition polymer of styrene, and accounts for approximately 10% of all plastics used in packaging in the U.S. Its structure is shown in Fig. 2.8.

Table 2.2 Typical Properties of Polypropylene and Polystyrene [1, 4, 5, 17, 20]

Property	PP	PS
T_g, °C (°F)	-10 (14)	74-105 (165-221)
T_m, °C (°F)	160-175 (320-347)	
Density, g/cm³ (lb/ft³)	0.89-0.91 (55.5-56.8)	1.04-1.05 (64.9-65.5)
Typical yield, m²/kg, 25 µm film (in²/lb, 1 mil)	44.0 (31,000)	36.9 (26,000)
Tensile strength, MPa (psi)	31-41.3 (4,500-6,000)	35.8-51.7(5,200-7,500)
Tensile modulus, MPa (10³ psi)	1,140-1,550 (165-225)	2,270-3,270(330-475)
Elongation at break, %	100-600	1.2-2.5
Tear strength, g/25 µm (lb/mil), film	50 (0.11)	4-20 (0.01-0.04)
WVTR, g µm/m² day at 37.8 °C, 90% RH (g mil/100 in² 24 h at 100 °F, 90% RH)	100-300 (0.25-0.76)	1,750-3,900 (4.4-10)
O₂ permeability, 25 °C,10⁴ cm³µm/m² day atm (77 °F, cm³ mil/100 in² 24 h atm)	5.0-9.4 (130-240)	9.8-15 (250-380)
CO₂ permeability, 25 °C,10⁴ cm³µm/m² day atm (77 °F, cm³ mil/100 in² 24 h atm)	20-32 (500-810)	35 (900)
Water absorption, %, 0.32 cm (125 mil) thick, 24 h	0.01-0.03	0.01-0.03

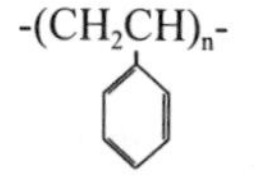

$$-(CH_2CH)_n-$$

Figure 2.8   Polystyrene

When viewed in the fully extended chain conformation, the benzene rings (C_6H_5 groups) on the main chain are ordered randomly (atactic PS). Consequently, polystyrene is unable to crystallize, and is an amorphous polymer. The bulky nature of the pendant group causes considerable resistance to rotation of the chain, leading PS to be a highly stiff, brittle material. Its lack of crystallinity makes it highly transparent. It is not suitable for use at high temperatures, as it experiences liquid flow at about 100 °C (212 °F). (Note amorphous polymers do not have a defined melting point, but gradually soften through a wide range of temperatures. For crystalline polymers, the melting point is defined as the relatively narrow temperature range over which the crystallites break up, leading to a discontinuous change in properties of the material.)

On the other hand, the tendency of PS to flow under stress at moderately elevated temperatures makes it easy to extrude and thermoform. Its brittleness can be reduced in two major ways. One is to incorporate a butadiene rubber fraction, partially as a graft copolymer and partially as a blend, to produce an impact grade of PS. The improved impact resistance brings a loss of transparency. Transparent grades, without impact modifiers, are generally referred to as crystal PS.

An even more widely used method to modify PS brittleness is to use it in a foamed form. In fact, PS foam is the most widely used packaging foam, and is an excellent cushioning and insulating material. The blowing agent used to produce the foaming action is typically either a hydrocarbon or carbon dioxide. Chlorofluorocarbons (CFCs) were used in production of some types of PS foams in the past, but this practice was discontinued in the U.S. several years ago and is now illegal in all developed countries, as a result of the Montreal Protocol agreement of 1989 on discontinuing use of ozone-depleting chemicals, and its revisions.

The brittleness of PS can also be significantly reduced by biaxial orientation of the polystyrene sheet or film. A variety of styrene-based copolymers have been developed to exhibit special combinations of properties.

Polystyrene is, in general, a low cost polymer, and a relatively poor barrier to water vapor and gases. Its chemical reactivity is greater than that of PE and PP. Polystyrene properties are summarized in Table 2.2.

2.5   Polyvinyl Chloride [1, 4, 5, 23, 24, 27-30]

Polyvinyl chloride (PVC) is an addition polymer of vinyl chloride, and also accounts for about 10 percent of all plastics used in packaging in the U.S. Its chemical formula is shown in Figure 2.9.

$$-(CH_2CH)-$$
$$|$$
$$Cl$$

Figure 2.9   Polyvinyl chloride

PVC is a partially syndiotactic polymer, crystalline, but to a very small extent. It has strong attractions between neighboring molecules, because of the polarity of the C-Cl bond, and in its normal state is stiff and rigid at room temperature. The melting temperature and decomposition temperature of PVC are very close together, rendering unmodified PVC very difficult to process. Decomposition produces HCl, which is highly corrosive, especially in the presence of water. To reduce problems associated with decomposition, stabilizers are added to PVC; octyl tins are most often used in rigid PVC for food and pharmaceutical packaging.

Plasticizers are often used to modify PVC, acting like internal lubricants, increasing the flexibility of the material. They also produce flow at lower temperatures, thereby decreasing processing temperatures. Flexible PVC, for packaging as well as other uses, is often highly plasticized. Many types of plasticizers are available, some suitable for food contact applications. Impact modifiers, such as methacrylate-butadiene-styrene (MPS), acrylonitrile-butadiene-styrene (ABS), chlorinated polyethylene, and acrylics are also blended with PVC.

The polar nature of PVC gives it a strong affinity for plasticizers as well as other additives. As a result, PVC can be produced with a wide stiffness range, from rigid containers to very soft and flexible films. Virtually all properties of PVC are strongly affected by the amounts and types of additives incorporated in its formulation. Relatively unplasticized resin PVC has reasonably good barrier properties, for example, while highly plasticized films provide poor barriers. Resin characteristics can also be modified by copolymerization. PVC properties are summarized in Table 2.3.

Table 2.3 Typical Properties of Polyvinyl Chloride and Polyethylene Terephthalate [1, 4, 5, 21-25]

Property	PVC	PET
T_g, °C (°F)	75-105 (167-221)	73-80 (163-176)
T_m, °C (°F)	212 (414)	245-265 (473-509)
Density, g/cm³ (lb/ft³)	1.35-1.41 (84.2-88.0)	1.29-1.40 (80.5-87.4)
Typical yield, m²/kg, 25 μm film (in²/lb, 1 mil)	27-30.5 (19,000-21,500)	30 (21,100)
Tensile strength, MPa (10^3 psi)	10.3-55.3 (1.49-8.02)	48.2-72.3 (7.0-10.5)
Tensile modulus, MPa (10^3 psi)	to 4,139 (600)	2,756-4,135 (400-600)
Elongation at break, %	14-450	30-3,000
Tear strength, g/25 μm (lb/mil), film		30 (0.066)
WVTR, g μm/m² day at 37.8°C, 90% RH (g mil/100 in² 24 h at 100 °F, 90% RH)	750-15,700 (1.9-40)	390-510 (1.0-1.3)
O_2 permeability, 25°C, 10^4 cm³μm/m² day atm (77 °F, cm³ mil/100 in² 24 h atm)	0.37-23.6 (9.4-600)	0.12-0.24 (3.0-6.1)
CO_2 permeability, 25°C, 10^4 cm³μm/m² day atm (77 °F, cm³ mil/100 in² 24 h atm)	1.1-19.7 (28-500)	0.59-0.98 (15-25)
Water absorption, %, 0.32 cm (125 mil) thick, 24 h	0.04-0.75	0.1-0.2

In appearance, PVC has good transparency, with a slight bluish tint. It yellows with age and therefore is often tinted a more pronounced blue, as this color is best at masking yellow. A primary use of PVC in packaging is for thermoformed blisters, although it is also used in water bottles, stretch wrap for meat packaging, and other applications. PVC bottles have been used much more widely in Europe than in the U.S., although this has decreased substantially because of environmental concerns.

One environmental issue related to PVC is the level of residual vinyl chloride monomer in the material, which may migrate into food. Vinyl chloride monomer has been determined to be a carcinogen, at least under some conditions. PVC packaging resins currently produced have much lower levels of residual vinyl chloride (under 10 ppb) than those used in containers in the mid-1970s when this concern first surfaced.

The disposal of PVC, especially when incinerated, raises other environmental questions. It is known that burning PVC produces HCl, and some suspect it contributes to the formation of chlorinated dioxins. These issues, along with more general concerns about the effects of chlorinated organics in our environment, have resulted in a rather negative environmental image for PVC, and has accelerated its replacement by PET and other plastics which provide some of the same functions without the perceived environmental side-effects.

2.6 Polyesters

Polyesters are a class of polymers containing ester linkages, with the general formula shown in Figure 2.10. Polyesters can be either thermoplastics or thermosets, depending on their chemical composition. By far the most commonly used polyester, both in packaging and other markets, is polyethylene terephthalate (PET). In fact, even polyester clothing and carpets are made of PET.

$$-O-(C-R-C-O-R'-O)_n-$$
$$\begin{array}{cc} \| & \| \\ O & O \end{array}$$

Figure 2.10   Polyester

2.6.1 Polyethylene Terephthalate [1 ,4, 5, 22, 25]

PET is a condensation polymer accounting for more than 10% of the plastic used in packaging in the U.S. While PET films have been available since the 1950s, the use of PET has expanded rapidly since its introduction in the mid-1970s as carbonated soft drink bottles. The structure of PET is shown in Figure 2.10. Raw materials for PET production are para-xylene and ethylene. The p-xylene is converted into either dimethyl terephthalate or terephthalic acid, and the ethylene into ethylene glycol. These monomers are then polymerized to form PET, using a condensation process that produces water as the byproduct molecule if terephthalic acid is used, and methanol as the byproduct if dimethyl terephthalate is used. The size of the polymer molecules, and hence the viscosity of the PET, is increased by following the condensation polymerization by a process known as "solid-stating," in which the dried and crystallized resin chips from the original polymerization are subjected to high temperatures.

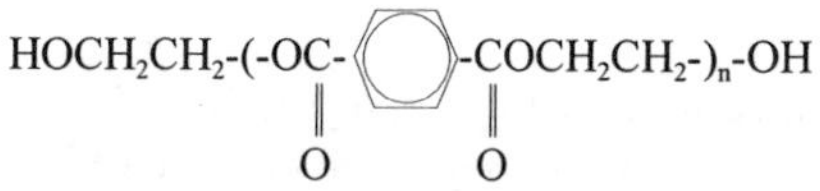

$$HOCH_2CH_2\text{-(-OC-}\bigcirc\text{-COCH}_2CH_2\text{-)}_n\text{-OH}$$

Figure 2.11   Polyethylene terephthalate

PET provides considerably better oxygen and carbon dioxide resistance than the plastics discussed so far in this book. The improved carbon dioxide barrier that results when PET is biaxially oriented opened the soft drink bottle market to this plastic, largely at the expense of glass. In recent years, PET has significantly expanded beyond beverage bottles into the "custom bottle" segment. Properties of PET are summarized in Table 2.3.

Biaxially oriented PET film has numerous packaging and non-packaging applications. Its odor barrier properties are excellent, and if additional barrier is required, PET film can be coated with polyvinylidene chloride (see Section 2.7.1), metallized with aluminum, or coated with silicon oxide (see Section 3.4).

While PET can crystallize to a high extent, it has a limited temperature range over which crystallization can occur. PET bottles and film are largely amorphous (APET), with small crystallites, and excellent transparency. However, crystallized PET (CPET) containers have a higher degree of crystallinity, larger crystallites, and are an opaque white. The more crystalline PET is much less subject to deformation under stress, especially at elevated temperatures, than its amorphous relative. It is also more brittle at cold temperatures. Copolymerization can be used to decrease the tendency of the material to crystallize.

CPET is used to produce ovenable trays for frozen food that can be heated either in a microwave or in a conventional oven. CPET's increased crystallinity greatly reduces tray deformation at elevated temperatures. In some instances, such trays have been made with a two-layer structure, with CPET for rigidity and APET for low temperature impact strength.

One of the disadvantages of PET is its low melt strength, which makes standard grades difficult or impossible to process by extrusion blow molding. Some specialty grades, produced by copolymerization or by increasing the molecular weight of the material, have improved melt strength, and can be formed by standard extrusion blow molding equipment.

PET is considerably less chemically stable than the addition polymers discussed so far in this book. In particular, PET is subject to hydrolysis at elevated temperatures, so pelletized PET must be dry before processing.

A significant factor in the growth of PET, particularly in the container market, is its perception as an "environmentally friendly" material. The existence of deposit legislation for soft drink bottles and the higher value and increased performance of PET compared to HDPE combined to give it a head start in efforts to establish plastics recycling. Recycled PET found an early market in polyester fiberfill, and continues to be the plastic with the highest recycling rate in the U.S.

2.6.2 Glycol Modified PET (PETG)

The properties of PET can be modified by copolymerizing it with an additional glycol, an additional diacid, or both. These additions generally reduce PET's crystallinity and increase its melt strength, permitting easier formability. Glycol modified PET (PETG) is the copolyester resulting from these reactions most often used in packaging.

PETG is a copolymer of cyclohexane dimethanol with ethylene glycol and terephthalic acid, characterized by high stiffness and hardness and good toughness even at low temperatures. It is amorphous, and remains clear and colorless even when processed into heavy sections. PETG's improved melt strength over PET allows it to be processed by injection molding, blow molding, or extrusion of shapes, tubing, film, or sheet. Uses of PETG include bottles for household and food products, blister packages, and medical device packaging. It is sterilizable by both ethylene oxide and gamma radiation.

2.6.3 Polyethylene Naphthalate (PEN) [26-29]

Polyethylene naphthalate is a condensation polymer of ethylene glycol and naphthalate dicarboxylate (NDC), with the structure illustrated in Fig. 2.12. Like PET, this polyester can crystallize, with the amount of crystallinity dependent on the processing history.

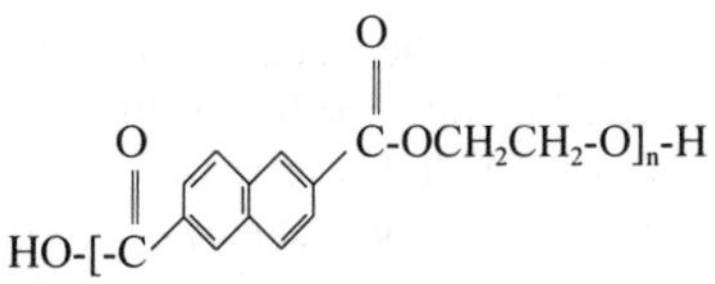

Figure 2.12   Polyethylene naphthalate

PEN has considerably better properties than PET in several respects, including a 400-500% better oxygen and water vapor barrier, a 35% higher tensile strength, and a 50% higher flexural modulus. It has better chemical resistance than PET, including greater resistance to hydrolysis, and has high resistance to UV-induced degradation, as well as the ability to protect its contents by blocking passage of UV light. Its improved high temperature performance permits hot filling without the side wall distortion which is a problem for PET. In addition, molding and blowing cycles are shorter than for PET, allowing for increased productivity. Of course, this comes at a price. PEN prices are estimated to remain at least three to four times those of PET for the forseeable future.

PEN homopolymer was approved by the FDA for food contact applications in April, 1996. PEN bottles have been recommended as refillable beverage bottles, since they can be subjected to higher temperature washing with caustic detergents than PET, and have a longer usable life.

PEN and PET can be blended together to make useful materials, and also PEN/PET copolymers can be produced. Because of the cost of PEN, this may open up larger markets than PEN alone. While these materials have not yet been approved for food contact applications in the U.S., such approval is anticipated.

Copolymers of PEN and PET are divided into two groups. Those containing less than 15% NDC are called low-NDC copolymers, while those containing 85% or more NDC are called high-NDC copolymers. Copolymers containing NDC levels between 15 and 85% do not currently have applications because of their amorphous nature. Homopolymer PET and PEN are immiscible, so blends of these materials require special dispersive mixing techniques. Blends of homopolymers with copolymers are easier to process. Typically a low-NDC copolymer is blended with PET, and a high-NDC copolymer with PEN, in efforts to achieve the right mix of properties while reducing costs.

2.7 Other Packaging Plastics

The plastics discussed above account for more than 95%, by weight, of all plastics used in packaging. However, a large number of other plastic resins are used for a variety of specialty applications. Some of these are discussed below.

2.7.1 Polyvinylidene Chloride [1, 4, 30]

Polyvinylidene chloride (PVDC) is an addition polymer of vinylidene chloride, with the structure shown in Fig. 2.13. In contrast to PVC, PVDC is highly crystalline. This, coupled with its tendency to decompose at elevated temperatures, like PVC, makes it extremely difficult to process, so it is always used in some modified form. Since the primary advantage of PVDC in packaging is its excellent barrier characteristics, the goal of such modification is to preserve the barrier, while improving processing characteristics. While small amounts of

$$-(CH_2CCl_2)_n-$$

Figure 2.13   Polyvinylidene chloride

plasticizers and other processing aids are often used, copolymerization is the major modification. Incorporation of a comonomer reduces crystallinity and the crystalline melting point, permitting processing at lower temperatures, or imparting solubility in organic solvents. Vinyl chloride and methyl acrylate are commonly used as comonomers for extrudable resins, typically in amounts from 6 to 28%. Vinylidene chloride copolymers with methyl acrylate and methyl methacrylate are commonly used for latex (water-based) coatings. Copolymers with acrylonitrile, methacrylonitrile, and methyl methacrylate are common for solvent-based coatings. Properties of PVDC are summarized in Table 2.4.

Table 2.4 Typical Properties of Polyvinylidene Chloride Copolymers and Ethylene Vinyl Alcohol [1, 4, 30, 31]

Property	PVDC	EVOH
T_g, °C (°F)	-15 to +2 (5-36)	48-69 (118-156)
T_m, °C (°F)	160-172 (320-342)	156-189 (313-372)
Density, g/cm³ (lb/ft³)	1.60-1.75 (100-109)	1.13-1.21 (70.5-75.5)
Typical yield, m²/kg, 25 μm film (in²/lb, 1 mil)	23.4 (16,500)	35 (25,000)
Tensile strength, MPa (10^3 psi)	19.3-34.5 (2.8-5.0)	37.2-94.1 (5.4-13.6)
Tensile modulus, MPa (10^3 psi)	344-551 (50-80)	
Elongation at break, %	160-400	180-330
Tear strength, g/25μm (lb/mil), film	10-30 (0.022-0.066)	
WVTR, g μm/m² day at 37.8°C, 90% RH (g mil/100 in² 24 h at 100 °F, 90% RH)	7.9-240 (0.02-0.61)	550-15,000 (1.40-38.1)
O_2 permeability, 25°C, cm³μm/m² day atm (77 °F, cm³ mil/100 in² 24 h atm)	7.9-2,700 (0.02-6.9)	2.6+ (0.0067+) (at 0% RH)
CO_2 permeability, 25°C, cm³μm/m² day atm (77 °F, cm³ mil/100 in² 24 h atm)	1,250-17,300 (3.2-44)	
Water absorption, %, 0.32 cm (1.25 mil) thick, 24 h	0.1	6.7-8.6

PVDC provides an excellent barrier to oxygen and other gases, as well as to water vapor, grease, odors and flavors, and its barrier capability is not affected by moisture. In addition, it has excellent ESCR and the ability to withstand hot-filling and retorting. PVDC copolymers are used in a variety of configurations in packaging, including monolayer films, but most often as part of a multilayer structure, either as a coating or as an inner layer in a coextruded structure.

In appearance, PVDC is highly transparent, with a yellowish tinge. It forms a very soft film, with excellent strength and self-cling. It has low melt strength, which requires special attention if it is being extruded as a monolayer material.

PVDC receives some of the same attention concerning the environmental impact of organochlorines as PVC. Proposed legislation to limit use of PVC sometimes includes PVDC as well.

2.7.2 Vinyl Acetate Polymers

2.7.2.1 Polyvinyl Acetate [4, 32]

Polyvinyl acetate (PVA) is an addition polymer of vinyl acetate, with the structure shown in Fig. 2.14. It is an atactic, amorphous polymer, tough and relatively stiff at room temperature, but tending to soften and become sticky when the temperature is only slightly elevated. The primary packaging use for polyvinyl acetate is in adhesive formulations, where the polar nature of the acetate group enhances its ability to bond substrates together, especially paper. This polarity also makes the properties change somewhat with changes in relative humidity and moisture content.

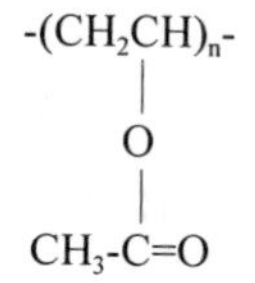

$$-(CH_2CH)_n-$$
$$|$$
$$O$$
$$|$$
$$CH_3-C=O$$

Figure 2.14   Polyvinyl acetate

2.7.2.2 Ethylene Vinyl Acetate [1, 3-5, 32, 33]

Ethylene vinyl acetate (EVA) is a copolymer of ethylene and vinyl acetate. Formulations vary widely in the relative fractions of ethylene and vinyl acetate they contain.

EVA resins used for films typically contain 5% vinyl acetate or less, to improve toughness and clarity. Even tougher films can be produced using 6 to 12% vinyl acetate. Resins designed for use as heat-seal layers in coextrusions often contain 15 to 18% vinyl acetate. These polymers are generally prepared under the high temperature/high pressure reaction conditions used for LDPE production, and as a result have a branched copolymer structure.

EVA copolymers are also used in adhesive formulations. The addition of some ethylene to the vinyl acetate improves adhesion of white glue formulations to plastics, in particular. EVA is often the major component in hot melt adhesives. These resins have a high percentage of vinyl acetate, with a low percentage of ethylene.

2.7.3 Vinyl Alcohol Polymers

2.7.3.1 Polyvinyl Alcohol

Polyvinyl alcohol (PVOH) has the following structure shown in Fig. 2.15. While it looks like an addition polymer of vinyl alcohol, in reality the vinyl alcohol monomer is unstable and PVOH is prepared by hydrolysis of PVA. The resulting material is amorphous at first, but tends to crystallize when oriented. It provides excellent barrier properties, because of its crystallinity and strong attractive forces between neighboring molecules. However, PVOH cannot be melt-processed since its decomposition temperature is below its melting temperature. In its pure form, PVOH is water-soluble, and the degree of solubility can be modified by controlling the amount of hydrolysis of the PVA. The more acetate groups that are retained, the lower the water solubility of the PVOH, as well as the less effective the barrier properties.

$$-(CH_2CH)_n-$$
$$|$$
$$OH$$

Figure 2.15   Polyvinyl alcohol

PVOH has few packaging applications because of its extreme water sensitivity and difficulty in processing. Important niche markets include packaging for toxic chemicals, such as herbicides, which can be encapsulated in PVOH pouches and placed, package and all, in the mixing tank. The PVOH dissolves, freeing the chemical, which can then be sprayed as usual. PVOH biodegrades in the environment, and does not generally clog spray nozzles or otherwise interfere with chemical application. In hospital laundries, PVOH bags of soiled linen can be placed directly in the washer, and hospital personnel avoid handling contaminated items.

2.7.3.2 Ethylene Vinyl Alcohol [1, 31]

The use of ethylene vinyl alcohol (EVOH) is rapidly expanding where excellent oxygen barrier is required. If EVA is hydrolyzed to EVOH, the resulting copolymer is no longer water soluble like polyvinyl alcohol, but retains excellent barrier capabilities. Formulations currently in use typically contain 27 to 48 mole % ethylene. The lower the amount of ethylene, the better the barrier when the polymer is dry, but the greater the moisture sensitivity. EVOH copolymers are highly crystalline, because of the ability of the -H and the -OH groups to fit into the same spot in the crystal lattice. They are also melt processable. EVOH is highly resistant to oils and organic solvents, and exhibits high strength, toughness, and clarity. For these reasons, EVOH resins are often the best choice for applications where excellent gas barrier is required, as long as exposure to moisture can be controlled. They are also very good as barriers to solvents, odors and flavors.

Because of its moisture sensitivity, EVOH is usually incorporated into package structures as a buried inner layer in a coextrusion, surrounded by polyolefins or other good water vapor barrier polymers. These structures typically contain an adhesive, or tie layer, between the EVOH and the polyolefin to provide adequate adhesion between the polar EVOH and the nonpolar polyolefin. For retorted products, this level of moisture protection may not be enough to prevent unacceptable oxygen permeation during processing. In that case, it is possible to incorporate a desiccant in the tie layer between the EVOH and the polyolefin. The desiccant then absorbs moisture which penetrates the polyolefin during retorting, preserving the dryness and hence the oxygen barrier of the EVOH. EVOH properties are summarized in Table 2.4.

2.7.4 Nylons (Polyamides) [1, 4, 5, 34]

Nylons, or polyamides, are a family of condensation polymers, generally linear, made from monomers with amine and carboxylic acid functional groups, resulting in amide linkages. The general structure is shown in Fig. 2.16.

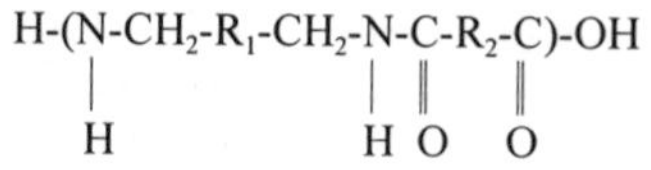

$$\text{H-(N-CH}_2\text{-R}_1\text{-CH}_2\text{-N-C-R}_2\text{-C)-OH}$$

$$\begin{array}{ccc} | & |\ || & || \\ \text{H} & \text{H O} & \text{O} \end{array}$$

Figure 2.16 Nylon (polyamide) structure

When nylons are polymerized from linear diamines and diacids, they are named according to the number of carbons in the diamine and diacid, respectively. Thus Nylon 6,10 is formed from a linear diamine containing 6 carbons and a linear diacid containing 10 carbons (Fig. 2.17).

Nylons polymerized from amino acids are given only one number, such as Nylon 11, corresponding to the number of carbons in the amino acids (Fig. 2.17). Most nylon packaging films used in the U.S. are Nylon 6, while in Europe they are most commonly Nylon 11.

Nylons can be characterized as semicrystalline materials with excellent mechanical properties and thermal stability. Their crystallinity is highly dependent on processing conditions, especially the rapidity of cooling. Their low temperature flexibility, yield strength, burst strength, and flex strength are all very good. They excel at providing a barrier to odors

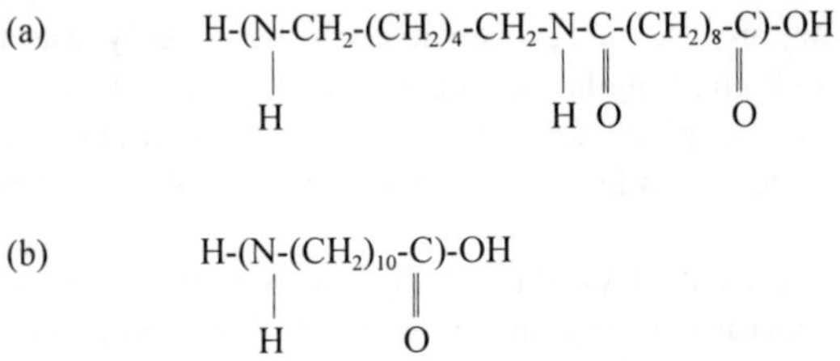

(a)   H-(N-CH$_2$-(CH$_2$)$_4$-CH$_2$-N-C-(CH$_2$)$_8$-C)-OH

       | | || ||

       H   H O   O

(b)   H-(N-(CH$_2$)$_{10}$-C)-OH

         | ||

         H   O

Figure 2.17   Typical nylons, (a) Nylon 6,10, (b) Nylon 11

and flavors, and provide good gas and oil barriers. The nylons' abilities to hydrogen-bond makes them quite moisture sensitive, and they do have some chemical reactivity. Nylons generally thermoform readily. They are often used in coextruded structures with other plastic materials, partly because of their relatively high cost. Typical properties of Nylon 6 and Nylon 11 are given in Table 2.5.

Table 2.5 Typical Properties of Nylon 6 and Nylon 11 [1, 4, 5, 34]

Property	Nylon 6	Nylon 11
T$_g$, °C (°F)	60 (140)	
T$_m$, °C (°F)	210-220 (410-428)	180-190 (356-374)
Density, g/cm^3 (lb/ft^3)	1.13-1.16 (70.5-72.4)	1.03-1.05 (64.3-65.5)
Typical yield, m^2/kg, 25 μm film (in^2/lb, 1 mil)	35 (25,000)	38.3 (27,000)
Tensile strength, MPa (10^3 psi)	41.3-165 (6.0-24)	55.1-65.4 (8.0-9.5)
Tensile modulus, MPa (10^3 psi)	689-1,700 (100-247)	1,270 (185)
Elongation at break, %	300	300-400
Tear strength, g/25 μm (lb/mil), film		400-500 (0.9-1.1)
WVTR, g μm/m^2 day at 37.8 °C, 90% RH (g mil/100 in^2 24 h at 100 °F, 90% RH)	3,900-4,300 (9.9-11)	1,000-2,000 (2.5-5.1)
O$_2$ permeability, 25 °C, cm^{3}μm/m^2 day atm (77 °F, cm^3 mil/100 in^2 24 h atm)	470-1,020 (1.2-2.6)	12,500 (32)
CO$_2$ permeability, 25 °C, cm^{3}μm/m^2 day atm (77 °F, cm^3 mil/100 in^2 24 h atm)	3,900-4,700 (10-12)	47,500 (120)
Water absorption, %, 0.32 cm (125 mil) thick, 24 h	1.3-1.9	0.4

In general, the longer the length of aliphatic chains (CH$_2$ groups) between -CONH- groups, the lower the melting point, tensile strength, and water absorption of the nylon and the greater its elongation and impact strength.

Packaging applications include oven cook-in bags, vacuum packages for processed meats, cheese packaging, modified atmosphere packaging, and medical packaging. Nylons can be sterilized with ethylene oxide or with steam, and some modified nylons can be radiation sterilized.

The properties of nylons, as of other polymers, can be modified by copolymerization. Variation of both the monomer(s) used and comonomers makes possible a wide range of properties. In general, because comonomers interfere with crystallization, copolymers will have lower melting points than the corresponding homopolymers.

Nylons formed from non-aliphatic diamines and/or diacids are becoming increasingly important in packaging. These nylons are often amorphous, and provide excellent clarity along with very good barrier properties, especially to oxygen.

MXD6 is a semicrystalline nylon produced by polymerization of meta-xylene diamine and adipic acid (Fig. 2.18). It is reported to have better gas barrier and thermal properties than Nylon 6, and better moisture resistance than EVOH. It has been used as a blend with PET, and has been approved by the FDA for some food contact applications. At 100% RH, the oxygen barrier of MXD6 is superior to that of EVOH. At low relative humidities, its barrier is inferior to PVDC at low temperatures, but superior at elevated temperatures. Odor and flavor barriers are also good, as are mechanical properties (see Table 2.6) [35].

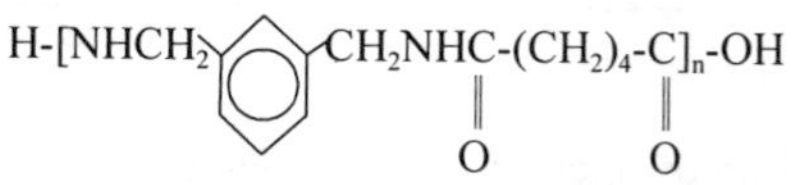

$$\text{H-[NHCH}_2\!\!\!\!\bigcirc\!\!\!\!\text{CH}_2\text{NHC-(CH}_2)_4\text{-C]}_n\text{-OH}$$

Figure 2.18   MXD6 nylon

Another nylon formed from nonaliphatic monomers is Selar PA, produced by DuPont, Inc. Selar PA is a copolymer of hexamethylene isophthalamide and terephthalamide, and therefore is totally amorphous. The gas barrier properties of Selar PA actually increase with increasing relative humidity, in sharp contrast to the behavior of other nylons. It has a glass transition temperature (T_g) of 127 °C (261 °F) and is typically melt-processed at 240-275 °C (464-527 °F). It is reported to have a much broader range of processing grades than semicrystalline nylons, along with excellent melt strength and very good oxygen, carbon dioxide, solvent, and water vapor barrier properties. Selar PA has excellent gloss, clarity, and high stiffness, giving it an appearance similar to glass in rigid containers [36].

Another use of nylon in packaging is in blends with polyethylene, under conditions in which the nylon forms "platelets" - small, plate-like regions of nylon embedded in the polyethylene matrix. These blends exhibit significantly increased barrier capabilities, particularly to hydrocarbons. This DuPont-patented technology involves drymixing Selar RB nylon barrier resin, typically 5 to 20% by weight, with polyolefins. The mixture can be processed in conventional blow molding equipment to produce containers with a series of discontinuous, two-dimensional, overlapping plates of nylon within the polyolefin matrix. The

Table 2.6 Typical Properties of MXD6 Nylon and Polycarbonate [35, 37]

Property	MXD6 (biaxially oriented)	PC
T_g, °C (°F)	64 (147)	150 (302)
T_m, °C (°F)	243 (470)	265 (510)
Density, g/cm³ (lb/ft³)	1.20-1.25 (75-78)	1.2 (75)
Typical yield, m²/kg, 25 μm film (in²/lb, 1 mil)	33 (23,200)	33 (23,200)
Tensile strength, MPa (10^3 psi)	220-230 (32-33)	63-72 (9.1-10.5)
Tensile modulus, MPa (10^3 psi)	3,820-4,110 (550-600)	2,380 (345)
Elongation at break, %	72-76	110-150
Tear strength, g/25 μm (lb/mil), film		10-16 (0.02-0.04)
WVTR, g μm/m² day at 37.8 °C, 90% RH (g mil/100 in² 24 h at 100 °F, 90% RH)	630 (1.6)	1,900-2,300 (4.9-5.9)
O_2 permeability, 25 °C, cm³μm/m² day atm (77 °F, cm³ mil/100 in² 24 h atm)	60-260 (0.15-0.66)	110,000 (300)
CO_2 permeability, 25 °C, cm³μm/m² day atm (77 °F, cm³ mil/100 in² 24 h atm)		675,000 (1700)
Water absorption, %, .32 cm (125 mil) thick, 24 h		0.15

result is up to a 140-fold improvement in hydrocarbon barrier compared to HDPE. Further, regrind can readily be recycled directly back into the process [38, 39].

2.7.5 Polycarbonate (PC) [1, 4, 37]

Polycarbonate is more properly known as poly(bisphenol-A carbonate) and has the structure shown in Fig. 2.19. It is an amorphous polymer, with excellent clarity and a very slight yellowish tinge. It is known for being very tough and rigid, with good impact strength, dimensional stability, and heat resistance, as well as its good performance at low temperatures. It has relatively poor chemical resistance to alkalis and some other chemicals, but good resistance to water, oil, and alcohols. PC thermoforms well, and is suitable for deep draws which may not be possible with some other materials. It can be sterilized by autoclaving or by electron beam or gamma radiation. The cost of PC is relatively high, and it provides a poor water and gas barrier. Properties of PC are summarized in Table 2.5.

Major uses of PC in packaging include medical packaging and 22.7 L (5-gal.) refillable water bottles, where it has wiped glass out of the market. Polycarbonate is also used for refillable milk bottles, although this application is rare. Because of the high price of PC, a 3.8 L (1 gal.) bottle usually carries a deposit of around 50 cents. That and the reluctance of retailers to handle return of the bottles has kept this market very small, even though it has been shown that the bottles can be refilled more than 50 times. In the last few years, several school systems

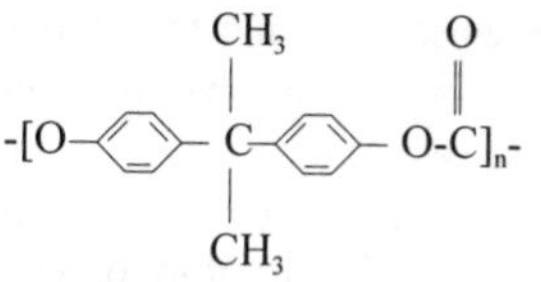

$$-[O-\langle\rangle-\underset{\underset{CH_3}{|}}{\overset{\overset{CH_3}{|}}{C}}-\langle\rangle-O-\overset{\overset{O}{\|}}{C}]_n-$$

Figure 2.19   Polycarbonate

have experimented with using single serving (0.25 L (1 cup)) bottles in school lunch programs as an alternative to cartons, as a waste (and hence cost) reduction measure. The reports are that the program has been successful in the few locations where it has been tried.

2.7.6 Ionomers [1, 4, 40]

Ionomers are polymers which contain interchain ionic bonding. They are typically sodium or zinc salts of ethylene/methacrylic acid copolymers or of ethylene/acrylic acid copolymers, polymerized at high temperatures and pressures as is LDPE, resulting in a branched structure. The general structure of the ethylene/methacrylic acid copolymers is shown in Fig. 2.20.

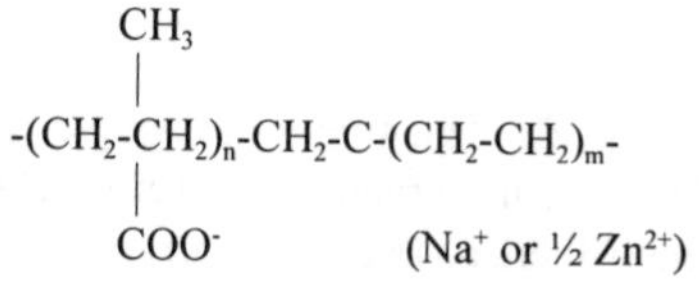

$$-(CH_2-CH_2)_n-CH_2-\underset{\underset{COO^-}{|}}{\overset{\overset{CH_3}{|}}{C}}-(CH_2-CH_2)_m- \qquad (Na^+ \text{ or } \tfrac{1}{2} Zn^{2+})$$

Figure 2.20   Ionomer

The ionic bonds function as reversible crosslinks, resulting in very strong attractions between neighboring molecules that can be weakened by heat, allowing the material to melt and flow. These links reformed on cooling. The amount of acid in the polymer backbone generally is between 7 and 30 weight %. Not all of the acid is neutralized to the salt. The amount of neutralization, typically in the 15 to 80% range, affects the ionomer's physical properties. The unneutralized acid provides sites for hydrogen bonding between adjacent molecules. These hydrogen bonds are weaker than the ionic bonds provided by the salt, but stronger than ordinary secondary bonds. Increasing acid content and percent neutralization increase tensile strength, modulus, toughness, clarity, melt strength, and oil resistance, while decreasing melting point and tear resistance. Zinc-based ionomers tend to have a wider range of adhesion and be less hygroscopic than sodium-based ionomers, while sodium ionomers have better optical properties and grease resistance.

Ionomers are also produced as terpolymers, containing isobutyl acrylate along with ethylene and methacrylic acid. Another family of ionomers is produced by saponification of ethylene-alkyl acrylate copolymers, with all the acid groups neutralized with sodium.

Ionomers have excellent melt strength, clarity, flexibility, strength and toughness. Their exceptional impact resistance and puncture resistance, even at low temperatures, makes them ideal for skin packaging of sharp objects. Their high infrared absorption, which leads to fast heating, is another advantage in skin packaging. Ionomers' excellent adhesion, and in particular their ability to heat seal very well, even through contamination, makes them widely used in vacuum packaging of processed meat. Their excellent melt strength is also useful in deep-draw thermoforming applications, and their superior hot tack makes them useful in vertical form/fill/seal packaging (see Section 3.6.4), where product is delivered into the package before the seal has completely cooled. Other uses of ionomers include the packaging of cheese, snack foods, and pharmaceuticals. Ionomers with unneutralized acid groups have excellent adhesion to aluminum foil and are often used as the heat-seal layer in extrusion-coated foil structures.

The disadvantages of ionomers include their relatively poor gas barrier, tendency to absorb moisture readily, and relatively high cost compared to PE and EVA.

2.7.7 Fluoropolymers [41]

Fluoropolymers contain C-F bonds. The simplest example is polytetrafluoroethylene (Fig. 2.21). Commonly available under the brand name Teflon (DuPont, Inc.), this very highly crystalline polymer is extremely inert, an excellent barrier, and exhibits a very low coefficient of friction. It is very difficult to process because of its very high viscosity. Its glass transition temperature (T_g) is about -100 °C (212 °F), and its melt temperature is about 327 °C (621 °F). It is also a high cost material, with primary uses in packaging as a component in packaging equipment, such as providing a non-stick surface on heat sealers, rather than in packages themselves.

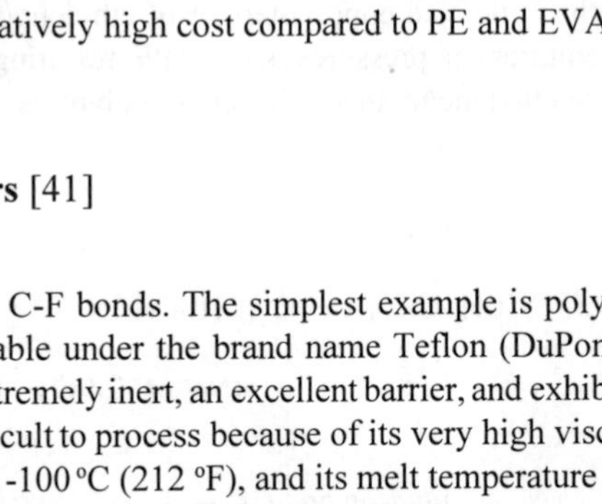

Figure 2.21 Fluoropolymers, (a) polytetrafluoroethylene, (b) polychlorotrifluoroethylene

Aclar film, from Allied Corp., is the only fluoropolymer film with significant packaging use. It is a modified (copolymerized) polychlorotrifluoroethylene (Fig. 2.21), which is available in several different grades. The polymer is semicrystalline, with a glass transition temperature about 45°C (113 °F) and a melt temperature of about 190 °C (374 °F) It can be melt processed, although with difficulty. Aclar's major advantage is its extremely good water vapor barrier, the best of any plastic film. Water vapor transmission rates at 38°C (100 °F) and 90% RH are 10-18 g μm/m^2 d (0.025-0.045 g mil/100 in^2 24 h). It also provides a good gas barrier, and is highly inert, although not particularly strong or tough. Oxygen permeability is about 2800-5900 cm^3 μm/m^2 d atm (7-15 cm^3 mil/100 in^2 24 h atm), and carbon dioxide permeability 6300-15,700 cm^3 μm/m^2 d atm (16-40 cm^3 mil/100 in^2 24 h atm). It can be used alone, but is most often used as a component in a laminated structure. Its major commercial application is for packaging moisture-sensitive drugs, but it also has military and other applications. Its cost is quite high [41].

2.8   Copolymers

Copolymers have already been mentioned a number of times. A copolymer can be defined simply as a polymer containing more than one type of repeating unit (monomer). Copolymerization is widely used to modify the properties of homopolymers (polymers containing only one kind of monomer) to achieve a desired set of properties.

2.8.1   Types of Copolymers

Copolymers can be categorized as random, alternating, block, and graft, depending on the arrangement of comonomers in their structure.

2.8.1.1 Random Copolymers

In random copolymers, there is no orderliness to the arrangement of the two (or more) different monomers (see Fig. 2.22). Introducing a different monomer reduces the structural regularity of the polymer. Thus, if one or both of the homopolymers is able to crystallize, the copolymer is expected to have less crystallinity, even at small levels of modification. At large levels of modification, the copolymer is likely to be totally amorphous, even if both homopolymers are highly crystalline. The exception to this general rule is when the monomers are isomorphous, meaning both can fit into the same spot in the crystal lattice. In that case, the presence of the comonomer does not interfere with crystallinity.

The properties of a polymer closely related to crystallinity are, of course, affected by the resultant decrease in crystallinity. Thus, copolymers tend to be more flexible and have lower tensile strength, but better impact strength and ESCR, than homopolymers.

Other properties of polymers are related less to crystallinity and more to the general strength of secondary bonds between molecules. For these properties, and especially for amorphous polymers, which do not have any crystallinity to diminish, the copolymer's properties tend to be a weighted average of the homopolymers' properties.

(a)   -AAABBABAABBBBBAABAAAAABBA-

(b)   -ABABABABABAB-

(c) $-(A)_n-(B)_m-(A)_p-(B)_q-$

(d) $-(A)_n-A-(A)_m-A-(A)_p-$
$$\qquad\quad | \qquad\quad |$$
$$\qquad (B)_q \quad (B)_r$$

Figure 2.22   Types of copolymers, (a) random, (b) alternating, (c) block, (d) graft.

2.8.1.2 Alternating Copolymers

In alternating copolymers, the order is as the name suggests, succeeding monomer units alternate in type (see Fig. 2.22). Alternating copolymers, in terms of properties, can be regarded as homopolymers with an -AB- repeating unit. Depending on their structure, they may or may not be crystalline, and have other properties also determined by that structure.

2.8.1.3 Block Copolymers

Block copolymers have long segments of one type of monomer followed by long segments of the other type of monomer (Fig. 2.22). The molecules tend to arrange themselves in small domains of a single type of monomer, so such polymers can have crystallites of polymer A and crystallites of polymer B, if both homopolymers can crystallize. Properties, similarly, tend to be a weighted average of the properties of the homopolymers. In many respects, these materials behave like polymer blends, but they do not have the problems caused by lack of adhesion between the two constituents often found in blends, since these constituents are joined to one another by multiple covalent bonds.

2.8.1.4 Graft Copolymers

Graft copolymers (Fig. 2.22) are quite similar to block copolymers, except that they consist of a main chain composed of one type of monomer, with branches composed of another type of monomer. The properties of graft copolymers are very much like those of block copolymers, with the same tendency for the materials to form microscopic domains of one type of monomer unit, and properties similar to blends.

2.8.2 Additional Packaging Copolymers

Many packaging copolymers have already been briefly discussed. A few of the other significant ones are discussed below.

2.8.2.1 Styrene-Butadiene Copolymers [1, 20, 42]

Block copolymers of styrene and butadiene are commonly used to add impact resistance to polystyrene. They are used in single-service food packaging, bottles, blister packs, overcaps, and film applications. They have good transparency and toughness, excellent shatter resistance, and are easily fabricated. The precise properties obtained depend, of course, on the ratio of the comonomers and the lengths of the blocks. In addition to other advantages, they are moderate-cost materials. They are widely used in medical packaging applications because they can be sterilized by both gamma irradiation and ethylene oxide.

Styrene-butadiene copolymers are often blended with other polymers. Transparent blends can be made with styrene, styrene-acrylonitrile copolymers, or styrene-methyl methacrylate copolymers. The addition of styrene results in significant loss of impact strength even at low levels, while certain blends of styrene-butadiene and styrene-methyl methacrylate have greatly improved impact strength. Nontransparent blends are made with high impact polystyrene, polypropylene, and polycarbonate.

Impact grades of polystyrene, specifically HIPS (High Impact PS), are partially graft copolymers and partially physical blends of polystyrene and polybutadiene. They typically contain 2 to 15 weight % polybutadiene. These materials are opaque, and in addition to significantly decreased brittleness, have a broad processing window and are easy to thermoform, either as sheet or as extruded foam.

2.8.2.2 Acrylonitrile Copolymers [1, 43]

Acrylonitrile polymers contain a carbon-nitrogen triple bond as a pendant group. The homopolymer, polyacrylonitrile, has the structure illustrated in Fig. 2.23. Because of the very

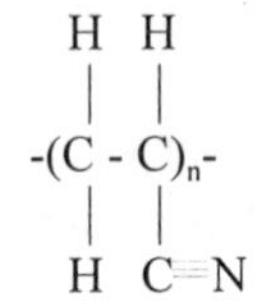

$$-(\overset{\displaystyle \overset{H}{|}}{\underset{\displaystyle \underset{H}{|}}{C}} - \overset{\displaystyle \overset{H}{|}}{\underset{\displaystyle \underset{C=N}{|}}{C})_n-$$

Figure 2.23   Polyacrylonitrile

strong polarity of the carbon-nitrogen bond, the polymer has extremely strong intermolecular forces, causing it to be an excellent gas barrier, but very stiff and brittle. It is not melt-processable, as it degrades at 220 °C (428 °F), below the temperature required for adequate flow. Thus packaging uses of acrylonitrile always employ a copolymer to impart melt-processability. Resins which are high in acrylonitrile, known as high-nitrile resins (HNR), are generally very tough materials, with good transparency and excellent barrier.

Styrene-acrylonitrile copolymer (SAN) is a random copolymer of styrene and acrylonitrile, most commonly made with a ratio of 3:1 styrene to acrylonitrile by weight. It is amorphous and transparent, with high heat resistance, excellent gloss and chemical resistance, and good tensile and flexural strength, and rigidity. It is used for cosmetic packaging, bottles, overcaps, closures, sprays, nozzles, and other applications. Gas barrier is fairly poor because of the low concentration of acrylonitrile groups. The precise polymer properties depend on the relative amounts of styrene and acrylonitrile present.

Acrylonitrile-styrene copolymers (ANS) generally contain a ratio of about 7:3 acrylonitrile to styrene by weight, and therefore have very good gas barrier properties. These resins were the first plastic bottles designed for carbonated beverages, but were removed from the market because of concerns over migration of acrylonitrile to the beverage. By the time the court case was settled and they were approved for that use (as long as residual acrylonitrile was under 0.1 ppm), PET had taken over the market.

Acrylonitrile-butadiene-styrene copolymers (ABS) are random styrene-acrylonitrile copolymers grafted to butadiene. These are amorphous materials, opaque, and easily processed. Again, the exact properties depend on the ratios of the comonomers used. ABS is used in cosmetics packaging, and has been used in margarine tubs (although most of these are now less expensive HDPE).

Copolymers containing acrylonitrile, methyl methacrylate, and rubber modifiers are also produced and used primarily in health care packaging, though they also have some food packaging use. These materials generally have only moderate gas barrier properties unless the acrylonitrile content is high. HNR with a 75:25 ratio of acrylonitrile to methacrylate polymerized onto a nitrile rubber backbone, produced by BP Chemicals under the trade name Barex, has excellent barrier properties. These resins can be used in the production of blow molded and injection-molded containers, film, and sheet, and are FDA-approved for direct food contact applications. In addition to excellent gas barrier, they have good sealability and chemical resistance. For medical packaging, they can be sterilized by either ethylene oxide or gamma radiation. Applications include rigid containers for spices, household chemicals, cosmetics, pesticides, agricultural chemicals, and fuel additives. Thermoformed blisters are used for meat and cheese packaging. Oxygen permeability for high nitrile resins is about 315-630 cm^3 μm/m^2 d kPa (0.8-1.6 cm^3 mil/100 in^2 24 h atm). Water vapor barrier is inferior to polyolefins, at about 2000-3000 g μm/m^2 d (5.0-7.5 g mil/100 in^2 24 h) at 38 °C (100 °F), 90% RH [43].

HNR is also used in coextruded structures, particularly with polyolefins. These can be in sheet, film, or bottle form. The HNR provides gas barrier and chemical resistance, while the polyolefin provides water vapor barrier and cost savings.

2.8.3 Tie Layers

When two different polymers are combined in a coextruded structure (see Chapters 3 and 5), there is often insufficient adhesion between them, especially if they are chemically dissimilar, such as where one is polar and one is non-polar. For this reason, these structures often require "tie layers," which function as adhesives.

Block and graft copolymers are often used in tie layer formulations. By providing domains containing chemical groups similar to both of the main polymers connected to each other by covalent bonds, they can effectively adhere to both of the major polymers, and thus unitize the structure.

2.9 Polymer Blends

Blends of polymers are sometimes used to obtain some of the advantages of both components, just as is done for copolymers. However, most polymers when blended together are mutually insoluble, resulting in the formation of small domains of one polymer within a matrix of another. Often this leads to unacceptable performance or performance dependent on the processing history, since this affects the morphology of the polymer blend. Nonetheless, there are applications for polymer blends in packaging.

Some applications of polymer blends have been discussed previously. LLDPE and LDPE are frequently blended to obtain the strength and puncture resistance of LLDPE with the improved processability and heat seal of LDPE. Impact resistant PS is partially a blend and partially a copolymer of styrene with butadiene. Blends of recycled polymers and virgin polymers are also utilized, and are discussed further in Chapter 9.

Blends of HDPE and LDPE, in a ratio of about 7:3, have been used for shrink films. The LDPE reduces the crystallinity of HDPE, permitting it to be stretched farther, while the crystallinity of HDPE gives increased tensile strength and shrink tension [44]. Blends of HDPE and LDPE are also used for grocery bags.

Styrene-butadiene copolymers are often blended with other resins. Blends of styrene-butadiene and polystyrene are used for sheet extrusion and thermoforming. Blends with styrene-acrylonitrile copolymers or styrene-methyl methacrylate copolymers are often used for injection molded applications. Blends with high impact polystyrene, polypropylene, and polycarbonate are also utilized [42].

A polymer blend with some unusual properties is HDPE with nylon. This material can be combined in an ordinary extruder to produce a blow molded container which has the nylon dispersed in the form of thin platelets in the HDPE matrix. This container has significantly enhanced barrier properties for gases, solvents, and odors. The nylon platelets disrupt the permeation of the gas through the HDPE matrix, causing the effective distance required for diffusion to be much greater. The containers' strength, toughness, and heat resistance are also improved. A typical container contains between 5 and 18% nylon by weight [39]. While these materials are useful for household chemicals and similar products, they have not been approved by the FDA for food contact applications.

2.10 Polymer Additives

Polymer resins for use in plastic packaging are never pure polymer. They always contain some type of additive mixture. Common types of additives include antioxidants and other stabilizers, lubricants, plasticizers, fillers, antistatic agents, and pigments, among others.

The primary function of lubricants is to prevent the polymer from sticking to molds or other processing devices during fabrication. This category of additives also includes viscosity depressants, which reduce the flow resistance of the polymer, and slip agents, which reduce the coefficient of friction. Common lubricants include fatty acid esters and amides, paraffin and polyethylene waxes, and metallic stearates such as zinc stearate. Calcium stearate is commonly used as an external mold-release agent. Antiblocking agents include organic amides and metallic soaps [45, 46].

Stabilizers prevent or retard decomposition of the polymer, primarily when it is exposed to heat during processing, or ultraviolet radiation during use. Virtually all polymers incorporate some type of antioxidant. Hindered phenols such as butylated hydroxytoluene (BHT) are commonly used for food packaging because they have been cleared by the U.S. FDA. Other antioxidants include organophosphites and thioesters, among many others. Heat stabilizers, such as barium-cadmium, organotin, lead, calcium-zinc, and antimony compounds, are commonly used with PVC to prevent generation of hydrogen chloride. There is decreasing use of lead stabilizers due to toxicity concerns. For food contact applications, octyltin mercaptide, calcium-zinc compounds, and methyltin have FDA clearance. Epoxidized soybean and linseed oil are often used as secondary additives to supplement these [45, 46]. UV stabilizers include phenyl salicylate, derivatives of 2-hydroxybenzophenone, zinc oxide, titanium dioxide, carbon black, nickel complexes and salts, hindered amines, and others. Hindered amine light stabilizers (HALS) have received FDA clearance, and are regarded as the most important type of UV stabilizers [45, 46].

Plasticizers are used especially with PVC to provide greater flexibility, softness, elongation, and low temperature strength. Most are high boiling point organic liquids, and include organic esters, hydrocarbon oils, non-polyester polymerics, and polar specialty types. Several hundred different plasticizers are available. Phthalic acid esters, such as dioctyl phthalate, are the most widely used, though use has been declining because of safety concerns. Dioctyl phthalate cannot be used in the U.S. for food contact applications, so citric acid esters, which have FDA clearance, are often substituted. Epoxidized oils, dioctyl adipate, and low molecular weight polyesters are also increasingly being substituted for dioctyl phthalate [45].

Antistatic agents are particularly important for polymers used with electronic equipment. Films containing these materials are often tinted pink for ease of recognition. Both internal and external antistatic agents can be used, with the mode of action the formation of a conducting layer which provides for the discharge of electrons. Common additives for internal antistatic agents are ethoxylated fatty amines and nonionic and quaternary ammonium compounds [45]. These migrate to the surface, a process called blooming, and cause the formation of a layer of conductive aqueous solution. In addition to use with electronic equipment, antistatic agents are used in other films to help prevent the attraction of dust particles.

Films used for applications such as fresh produce packaging, where droplets of moisture can condense and block the view of the product, are often treated with antifog agents. These agents, such as nonionic ethoxylates or hydrophilic fatty esters such as glyceryl stearate, tend to combine the moisture into a continuous film rather than droplets, thus enhancing product visibility [45].

Nucleating agents are used to speed up crystal growth in certain polymers such as polypropylene. By providing sites for crystal growth to start, they produce more rapid and more uniform crystallites, enhancing the transparency of the bottle or film. Adipic and benzoic acid and some of their metal salts are commonly used in polypropylene for this purpose. Colloidal silicas are used in nylon [46].

Fillers of various types can be incorporated into polymers. Inorganic minerals such as talc, clay, calcium carbonate, and titanium dioxide can be added to reduce resin cost, increase stiffness, and increase the temperature at which the polymers can be used. However they generally decrease elongation, tensile strength, and impact strength. Glass fibers and mica can be used to significantly improve stiffness, and in some cases to add strength. In general, fibrous fillers are more effective at improving properties than particulates. In all cases, adhesion between the filler and the polymer matrix is necessary for enhanced strength. Coupling agents such as titanates can be added to improve this adhesion [45, 47].

For some specialty applications, additives such as antimicrobials to prevent growth of fungi and bacteria, flame retardants, fragrance enhancers, and other additives are used. Plastics intended for use in foams often incorporate blowing agents. Another important category of additives is colorants, which are discussed in Chapter 7.

When formulating an additive package, it is important to know how one additive can influence the effects of others. The cumulative effect can be greater in the desired direction, in which case the additives are termed synergistic. On the other hand, the additives may be antagonistic, interfering with each other.

References

1. *Modern Plastics Encyclopedia Handbook* (1994) New York: McGraw-Hill
2. Smith, M.A. (1986) In *The Wiley Encyclopedia of Packaging Technology*, M. Bakker (Ed.), New York: John Wiley & Sons, pp. 514-523
3. Maraschin, N.J. (1997) In *The Wiley Encyclopedia of Packaging Technology, 2nd ed*, A.L. Brody and K.S. Marsh (Eds.), New York: John Wiley & Sons, pp. 752-758
4. Briston, J.H. (1989) *Plastic Films*, 3rd ed., Essex, England: Longman Scientific & Technical
5. Benning, C.J. (1983) *Plastic Films for Packaging: Technology, Applications and Process Economics*, Lancaster, Pa.: Technomic Pub. Co., Inc.
6. Brody, A.L. and Marsh, K.S. (1997) In *The Wiley Encyclopedia of Packaging Technology, 2nd ed*, A.L. Brody and K.S. Marsh (Eds.), New York: John Wiley & Sons, pp. 405-407
7. Van der Sanden, D. and Halle, R.W. (May 1993) Packaging Technology & Engineering, pp. 32-37
8. Simon, D.F. (April 1994) Packaging Technology & Engineering, pp. 34-37
9. Vernyi, B. (June 26, 1995) Plastics News, pp. 1, 14
10. Vernyi, B. (Oct. 2, 1995) Plastics News, pp. 6, 24
11. Leaversuch, R.D. (March, 1994) Modern Plastics, p. 86
12. Modern Plastics (July, 1994), pp. 33-34
13. Toensmeier, P.A. (Jan., 1995), Modern Plastics, pp. 17-18
14. Leaversuch, R. (Dec., 1995) Modern Plastics, pp. 71-73
15. Thayer, A.M. (Sept. 11, 1995) C&EN, pp. 15-20
16. Plastics News (June 26, 1995), p. 15
17. Miller, R.C. (1997) In *The Wiley Encyclopedia of Packaging Technology, 2nd ed*, A.L. Brody and K.S. Marsh (Eds.), New York: John Wiley & Sons, pp. 765-768
18. Kong, D. and Mount, E.M. (1997) In *The Wiley Encyclopedia of Packaging Technology, 2nd ed*, A.L. Brody and K. S. Marsh (Eds.), New York: John Wiley & Sons, pp. 407-408
19. Mount, E.M. III and Wagner, J.R. Jr. (1997) In *The Wiley Encyclopedia of Packaging Technology, 2nd ed*, A.L. Brody and K.S. Marsh (Eds.), New York: John Wiley & Sons, pp. 415-422
20. Wagner, P.A. and Sugden, J. (1997) In *The Wiley Encyclopedia of Packaging Technology, 2nd ed*, A.L. Brody and K.S. Marsh (Eds.), New York: John Wiley & Sons, pp. 768-771
21. Cocco, D.A. (1997) In *The Wiley Encyclopedia of Packaging Technology, 2nd ed*, A.L. Brody and K.S. Marsh (Eds.), New York: John Wiley & Sons, pp. 771-775
22. Brody, A.L. and Marsh, K.S. (1997) In *The Wiley Encyclopedia of Packaging Technology, 2nd ed*, A.L. Brody and K.S. Marsh (Eds.), New York: John Wiley & Sons, pp. 742-745
23. Doshi, A.G. (1997) In *The Wiley Encyclopedia of Packaging Technology, 2nd ed*, A.L. Brody and K.S. Marsh (Eds.), New York: John Wiley & Sons, pp. 401-403
24. van Beek, H.J.G. and Ryder, R.G. (1997) In *The Wiley Encyclopedia of Packaging Technology, 2nd ed*, A.L. Brody and K.S. Marsh (Eds.), New York: John Wiley & Sons, pp. 427-431

25. Newton, J. (1997) In *The Wiley Encyclopedia of Packaging Technology, 2nd ed*, A.L. Brody and K.S. Marsh (Eds.), New York: John Wiley & Sons, pp. 408-415
26. Ford, T. (Sept. 25,1995) Plastics News, p. 7
27. Bregar, B. (Nov. 13, 1995) Plastics News, pp. 15, 36
28. Reynolds, P. (July, 1995) Packaging World, pp. 26-30
29. Leaversuch, R.D. (Nov., 1995) Modern Plastics, pp. 19-21
30. DeLassus, P.T., Brown, W.E. and Howell, B.A. (1997) In *The Wiley Encyclopedia of Packaging Technology, 2nd ed*, A.L. Brody and K.S. Marsh (Eds.), New York: John Wiley & Sons, pp. 958-961
31. Foster, R. (1997) In *The Wiley Encyclopedia of Packaging Technology, 2nd ed*, A.L. Brody and K.S. Marsh (Eds.), New York: John Wiley & Sons, pp. 355-360
32. Kaye, I. (1997) In *The Wiley Encyclopedia of Packaging Technology, 2nd ed*, A.L. Brody and K.S. Marsh (Eds.), New York: John Wiley & Sons, pp. 23-25
33. Hatfield, E. and Horvath, L. (1997) In *The Wiley Encyclopedia of Packaging Technology, 2nd ed*, A.L. Brody and K.S. Marsh (Eds.), New York: John Wiley & Sons, pp. 237-240
34. Brody, A.L. and Marsh, K.S. (1997) In *The Wiley Encyclopedia of Packaging Technology, 2nd ed*, A.L. Brody and K.S. Marsh (Eds.), New York: John Wiley & Sons, pp. 681-686
35. Watanabe, H. (undated) *A New Super Gas Barrier Nylon for Packaging--MXD6*, Toyobo Co., Ltd., Japan
36. DuPont (undated) *Selar PA 3426 Barrier Resin*, Wilmington, Del.
37. Brody, A.L. and Marsh, K.S. (1997) In *The Wiley Encyclopedia of Packaging Technology, 2nd ed*, A.L. Brody and K.S. Marsh (Eds.), New York: John Wiley & Sons, pp. 740-742
38. DuPont (undated) *Selar Barrier Resin*, Wilmington, Del.
39. Brody, A.L. and Marsh, K.S. (1997) In *The Wiley Encyclopedia of Packaging Technology, 2nd ed*, A.L. Brody and K.S. Marsh (Eds.), New York: John Wiley & Sons, pp. 864-867
40. Hoh, G. (1997) In *The Wiley Encyclopedia of Packaging Technology, 2nd ed*, A.L. Brody and K.S. Marsh (Eds.), New York: John Wiley & Sons, pp. 527-529
41. Brody, A.L. and Marsh, K.S. (1997) In *The Wiley Encyclopedia of Packaging Technology, 2nd ed*, A.L. Brody and K.S. Marsh (Eds.), New York: John Wiley & Sons, pp. 403-405
42. Hartsock, D.L. (1997) In *The Wiley Encyclopedia of Packaging Technology, 2nd ed*, A.L. Brody and K.S. Marsh (Eds.), New York: John Wiley & Sons, pp. 863-864
43. Lund, P.R. and McCaul, J.P. (1997) In *The Wiley Encyclopedia of Packaging Technology, 2nd ed*, A.L. Brody and K.S. Marsh (Eds.), New York: John Wiley & Sons, pp. 669-672
44. Jolley, C.R. and Wofford, G.D. (1997) In *The Wiley Encyclopedia of Packaging Technology, 2nd ed*, A.L. Brody and K.S. Marsh (Eds.), New York: John Wiley & Sons, pp. 431-434
45. Seymour, R.B. (1986) In *The Wiley Encyclopedia of Packaging Technology*, M. Bakker (Ed.), New York: John Wiley & Sons, pp. 2-4
46. Rosato, D.V. (1997) In *The Wiley Encyclopedia of Packaging Technology, 2nd ed*, A.L. Brody and K.S. Marsh (Eds.), New York: John Wiley & Sons, pp. 8-13.
47. Throne, J.S. (1986) In *The Wiley Encyclopedia of Packaging Technology*, M. Bakker (Ed.), New York: John Wiley & Sons, pp. 529-536

3   Extrusion, Films, and Flexible Packaging

3.1   Extrusion [1-3]

In nearly all applications for packaging plastics, the first step is to convert the solid plastic, usually in pellet form, into a melt, which can then be shaped using heat and pressure. The equipment used to do this is an extruder for film, sheet, and extrusion blow molded bottles, and an injection molding machine for injection molded and injection blow molded packages and components. Extruders and injection molding machines are very similar in concept and structure, differing primarily in how the material leaves the chamber where it is melted.

The purpose of an extruder or an injection molding machine is to take the solid plastic and, using heat, pressure, and shear, transform it into a uniform melt which can then be delivered to the next processing stage. Producing the melt may involve mixing in additives such as color concentrates, blending resins together, and incorporating regrind. The final melt must be uniform both in concentration and in temperature. The pressure must be high enough to force the viscous polymer through an opening (die) which imparts a desired shape to the extrudate, or to force the melt into a mold chamber.

The extruder accomplishes all this by using a barrel containing a screw with helical channels. The plastic pellets are fed into the barrel through a hopper at one end, and then conveyed by the screw to the other end of the barrel. The depth of the channels in the screw decreases with distance from the hopper, building up pressure on the plastic. External heating as well as internal heating from friction between the plastic and the screw cause the plastic to soften and melt. A simplified extruder diagram is shown in Fig. 3.1. Extruder designs are frequently tailored to the needs of particular polymers and particular applications. Some of the many options are incorporation of venting, multiple feed ports, special mixing devices along the screw, provision for cooling as well as heating of the melt, or provision for no external heat source at all (an adiabatic extruder), relativeness abruptness in the change in clearance between

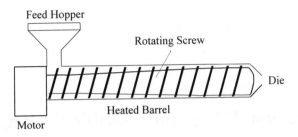

Figure 3.1   Plastics Extruder

the screw and barrel, and even the number of screws in the barrel. Twin screw extruders can provide for more intensive mixing than single screw extruders, for example. Tandem extrusion involves feeding a second extruder with the melt from the first extruder, and is commonly used in producing extruded polystyrene foam.

Extruders are sized by the diameter of the barrel and the ratio of the length of the barrel to the diameter. A typical extruder size for packaging applications would be a barrel diameter of 5 to 20 cm (2 to 8 inches) and a length to diameter (L/D) ratio of about 32:1. Extruders generally contain at least three sections. The first, next to the feed hopper, is the feed section. Its function is to get the plastic into the extruder at a relatively even rate. Generally, this section is maintained at a relatively cool temperature, to avoid blockage in the feed channel. Next comes the compression section, in which most of the melting occurs and pressure is built up. The transition between the feed section and the compression section can be abrupt or gradual. Finally, next to the extruder outlet, is the metering section. Its purpose is to assure a uniform flow of material (a "metered" flow) out of the extruder. Here sufficient residence time is needed to ensure that both composition and temperature are uniform.

At the end of the barrel, the melted plastic leaves the extruder through a die, which has been designed to impart the desired shape to the stream of melted plastic. This may be the final shape for the plastic, or it simply may be the beginning of further processing.

Another important part of the extruder is the drive mechanism. It controls the speed of rotation of the screw, which is the primary determinant of the output of the extruder. The amount of power required is affected by the viscosity (resistance to flow) of the polymer. Polymer viscosity is dependent on both temperature and flow rate, decreasing with increasing temperature and with increasing shear.

Extruders generally contain screens for filtering impurities out of the melt. Provisions may exist for automatic screen changes, to avoid downtime. This is particularly important when contaminated streams are being processed, such as recycled material.

3.2   Cast Film and Sheet [1-5]

For the production of cast film or sheet, the die opening through which the melted plastic exits the extruder is shaped like a slit, producing a rectangular profile in the melt, with the width much greater than the thickness. The two major types of cast film are cold cast film and quench tank cooled film.

Most commonly, cast film and sheet are produced by extrusion of the melt onto chilled chrome rollers (Fig. 3.2). The plastic exits the die downwards or at an angle between vertical and horizontal, onto the chill roll, contacting it tangentially. The roll is highly polished to give good surface characteristics to the resulting film. Generally, the film travels in an S-pattern around at least two chill rolls, sometimes more, before the film is cool enough to trim and wind. In fact, the first chill roll typically operates at a temperature of at least 40 °C (104 °F). Usually an air knife is used to pin the plastic to the chill roll. This method is used more often than the water bath method, described next, since it tends to produce film with better transparency and increased stiffness, and may also increase output. The film dimensions are controlled primarily by the extrusion rate and take-off speed.

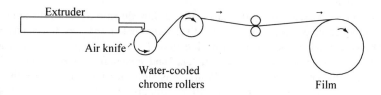

Extruder

Air knife

Water-cooled
chrome rollers

Film

Figure 3.2   Cast film using a chill roll

Occasionally, film or sheet is cast with the extrudate descending vertically into a quench tank of cooling water where it solidifies, and then dries and is rolled up (Fig. 3.3). Drying may be accomplished by evaporation alone, but is commonly aided by air jets, rolls, or radiant heat. The film characteristics are controlled by the die dimensions, extrusion rate, melt temperature, drawdown, and water temperature.

In either method, the plastic shrinks as it cools, so the film or sheet produced is narrower than the die dimensions, and also tends to thicken at the edges, which must be trimmed. Also, any irregularities in the gauge tend to be magnified when the film is rolled up. Cast film variations of ±3% are common, and can produce gauge bands in the roll which can cause difficulties in later converting operations. The problem of gauge bands can be minimized in two ways. The oldest method is to oscillate the film as it is wound to produce some randomization of the thickness variations. This method has the disadvantage of significantly

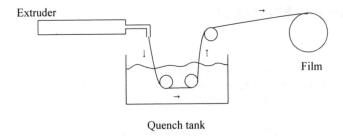

Figure 3.3   Cast film using a quench tank

reducing the width of the web and increasing the amount of scrap generated, which is generally fed back into the extruder in-line. A more modern method is to use sensors to monitor the thickness on-line, and via computer controls automatically adjust the die gaps as needed.

Cast film properties can be modified by orientation, or stretching of the film under conditions causing some molecular realignment in the direction of the stretch, which persists when the force is removed. Orientation tends to increase crystallinity, barrier properties, and strength in the stretched direction, while decreasing strength in directions perpendicular to the orientation direction.

If the cast film is not stretched significantly in the machine direction (the direction of travel through the production equipment), it is relatively unoriented and has fairly balanced mechanical properties in the machine and cross directions. If the takeoff speed is significantly higher than the rate of extrusion, the plastic is stretched and uniaxially oriented. The stretching can occur at the initial contact with the chill roll, but is more commonly done after the first chill roll, and usually involves some reheating of the film prior to stretching it. If the plastic is also stretched in the cross machine direction, it is biaxially oriented. Biaxial orientation can be accomplished in either a single step or, as occurs more commonly, in two consecutive steps (see Figs. 3.4 and 3.5). The resulting film is "balanced" if the orientation is equal in the two directions, or "unbalanced" if it is more highly oriented in one direction than the other. For effective orientation, a crystalline polymer must be below its melt temperature, but warm enough for some mobility of the molecules. The higher the orientation temperature, the more the material tends to flow, and the less actual orientation is produced. After drawing, the film is annealed to improve its thermal stability, if desired, and then cooled to "freeze" the orientation before the tension is released. It is also important that the crystallization be homogeneous and not excessive. Temperature control can be improved by placing the film in contact with oil-heated rolls, or other heating mechanisms.

Two-step orientation is widely used for PET film. The film is commonly drawn to three to four times its original length, and then to three to four times its original width. Oriented PP film (OPP) is also produced in this way, with varying degrees of orientation.

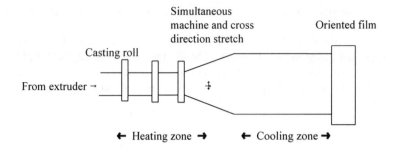

Figure 3.4   One-step orientation of cast film

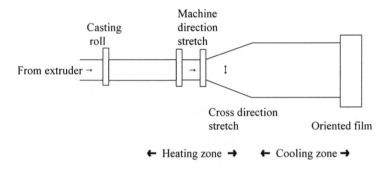

Figure 3.5   Two-step orientation of cast film

3.3   **Blown Film** [1-3,6]

Blown film is produced through an annular die, thus extruding a hollow tube of plastic often called a "bubble." Air pressure expands this tube into the desired dimensions. Usually the tube is extruded vertically upward, although it may be horizontal or even downward. Cooling is typically accomplished by blowing cool air on the film, usually through a cooling ring mounted directly on the die outlet. Internal as well as external cooling may be provided. Sometimes the tube is quickly cooled and then reheated before stretching with air pressure.

The dimensions of the film are controlled by the extruder output and the internal air pressure. The "blow-up ratio" is the ratio between the diameter of the final tube of film and that of the die. The internal air pressure is typically supplied through a port into the mandrel (the interior portion of the die), although it may be injected through the bubble wall, which then reseals itself. Generally once the process is running steadily, little air is lost, so make-up requirements are small.

The passage of the film through the tower is typically aided by various guiding devices, including air rings and guide rollers. When the film is sufficiently cooled, the bubble is collapsed by plates and rollers (pinch rollers), and wound up, with or without slitting, gusseting, or other treatment. Thus this process can be used to produce tubular film as well as flat film (Fig. 3.6).

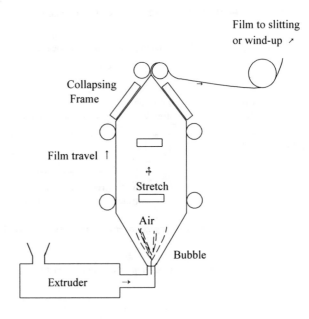

Figure 3.6   Production of blown film

Once the blown film process is up and running, little scrap is produced, since the edge effects associated with cast film are not present. However, getting the line running properly is more complex, so more scrap is produced during start-up. Blown film lines generally have higher output than cast lines, and are more economical once started up. Thus, blown film excels for high production runs. It is estimated that about 90% of all polyethylene film is produced this way [6].

Blown film tends to be of somewhat lower quality than cast film in terms of transparency and uniformity of gauge. The slower cooling of blown film is largely responsible. For transparency, slower cooling allows time for a greater degree of crystallinity and larger crystallites to develop, making the film hazier. The more uneven cooling contributes to greater variation in thickness. Since even a slight lack of perfect concentricity and freedom from defects in the die opening could lead to significant unevenness in the profile of the film roll, it is common practice to spin the die during extrusion, thus randomizing the thickness variation and producing a uniform roll of film. Variation in gauge can be as much as ±15% in some cases [6], though variations of ±7% are more common [2]. It is also possible to rotate the tube during winding, as an alternative to spinning the die.

In blown film, typically machine direction orientation is provided by stretching the film between the die and the pinch rolls. Orientation is also provided by the transverse stretching done as the bubble expands. Thus, the film is biaxially oriented, with the balance of the orientation produced by the relative amounts of stretching in the two directions. Additional orientation can be produced by further processing downstream of the blowing unit. Some blown film lines split the extrudate and feed two dies at once, each with its own blown film set-ups.

Oriented polypropylene film is generally extruded in a downwards direction. The extruded tube is cooled and then reheated, to a point still below its melt temperature, before it is blown. If a heat-stabilized film is desired, the collapsed bubble can be fed over a series of heated rollers, or it can be reinflated in what is known as the "double bubble" process. In either case, the film is reheated to relieve stresses and then cooled, while it is restrained from shrinking until cooling is complete.

3.4 Multi-Layer Film and Sheet

Many packages are produced from combinations of materials to optimize the needed performance by minimizing material usage and hence cost. This is especially true for flexible packaging materials, where often the ability to replace a rigid or semirigid package by a flexible one is directly dependent on the use of combinations of materials such as paper, foil, and/or plastic. Typically in this case the multilayer package gets most of its strength from a relatively inexpensive plastic or paper, and depends on a higher cost material for properties such as barrier. In other examples, multilayer materials are used to protect printing, provide different appearances or surface characteristics on the two sides of a material, or for a number of other purposes. Multilayer flexible packaging materials can be divided into three categories: coatings, laminations, and coextrusions.

3.4.1 Coatings

Coatings are produced, as the name indicates, by applying a thin layer of another material on one or both sides of the web, typically from a melt or from a solution or emulsion. The most common reasons for applying a coating are to add heat sealability to the base material, and to improve its barrier properties. Common heat-seal coatings are polyethylene and ethylene vinyl acetate. PVDC copolymers can be used for both barrier and heat-sealability.

Extrusion coating resembles cold cast film extrusion, except that the melted polymer is deposited on a web of film, paper, paperboard, or foil, rather than on a chrome roll. It is essential that there be enough compatibility between the coating and the substrate for good bonding.

3.4.1.1 Metallized Film [1,8-10]

A barrier material of particular interest is metallized film. In this process, the film to be metallized is placed in a vacuum chamber. Under very high vacuum, an aluminum wire is vaporized in a crucible, or boat, at about 1700 °C (3100 °F) and the aluminum condensed in an extremely thin layer on the film, typically 400 to 500 Å (Fig. 3.7). Heat is supplied by resistance heating, induction heating, or from an electron-beam source. The metal coating thickness is commonly specified in terms of light transmission, optical density, or electrical resistance. Coating rates can be up to 500 m/min. The film passes through several rollers to adjust tension and cool, and is then wound up on a take-up roll after the aluminum has been deposited. The thickness of the aluminum layer is controlled by the metal temperature (a higher temperature gives a thicker coating), running speed of the film, and number of plating stations. Coating an average sized roll takes about 30 minutes. Roll widths vary from 0.6 to 2.1 m (24 to 84 in.). Continuous metallizers have also been developed, but are not often used for packaging applications because of the high cost of the equipment; batch processes predominate.

The result of vacuum metallizing is a film with mechanical properties essentially identical to the base film, but with barrier properties nearly as good as aluminum foil. The flexibility of the film is maintained, avoiding the flex-cracking problems sometimes encountered with foil. In fact, when metallized foil is flexed and compared to foil laminations which have been flexed, the barrier of the metallized film can be superior to the foil. An oxygen barrier for 14 μm (0.5 mil) metallized polyester of less than 0.15 cc/m^2 MPa d (<0.01 cc/100 in^2 24 h atm), and water vapor transmission rate of less than 0.62 g/m^2 d (< 0.04 g/100 in^2 24 h) have been reported [11]. The attractive, shiny appearance of aluminum foil is also achieved. Overall costs are significantly less than for the much thicker laminations containing aluminum foil and paper, which metallized foil has replaced in many applications.

Common film substrates for metallizing include biaxially oriented nylon, PET, and PP. LDPE, PC, cellophane, and other films can also be metallized, but are used less frequently. Metallized paper is often used in packaging for its decorative apearance, but does not have very good barrier properties. Metallized films can also be used for their decorative appeal, but are much more significant as barrier materials. One disadvantage of metallized films is their opacity. Techniques exist for confining the metallization to certain areas of the film, so that transparent stripes are produced. Other shapes can be produced by removing some of the metal during the printing operation [9].

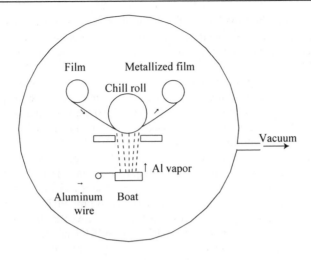

Figure 3.7 Metallization of film

3.4.1.2 Silicon Oxide Coated Film [12-15]

More recently, a similar idea has been applied to produce another type of high-barrier flexible package that has an advantage that metallized foil does not - it can be microwaved. If a thin layer of silicon oxide, similar to glass, is deposited on a flexible package, the result is also a material with excellent barrier capability that largely maintains the mechanical properties of the base film. The SiO_x layer sometimes imparts a slight yellowish color to the film, but preserves its transparency to microwave radiation. PET is the usual base substrate for SiO_x coated films, in 13-25 μm (0.5-1.0 mil) thicknesses, but polypropylene, polystyrene, and polyamides may also be used. The coating can be deposited by evaporation, sputtering, or the use of a chemical plasma.

The evaporation process is very similar in concept to vacuum metallizing, except that the energy for vaporization is typically delivered by a high energy electron beam. The SiO_x thickness is typically 400 to 1000 Å, with the number of oxygen atoms per silicon atom averaging between 1.0 and 2.0. The vacuum used must be several orders of magnitude greater than that used for aluminum metallization. Only about 20% of the SiO_x emitted from the source is deposited on the film, with the remainder deposited on the interior of the chamber. PET/PP laminates with an interior layer of SiO_x produced by electron beam evaporation have shown a reduction in oxygen permeation from 112 cm^3/m^2 d (7.2 $cm^3/100$ in^2 day) to 0.47 cm^3/m^2 d (0.03 $cm^3/100$ in^2 day), and water vapor permeation from 1.1 g/m^2 d (0.07 $g/100$ in^2 d) to 0.47 g/m^2 d (0.03 $g/100$ in^2) [12]. The precise barrier obtained depends on the base film as well as the coating conditions and method. Significant yellowing of the film usually occurs because of the presence of relatively high amounts of silicon monoxide.

The sputtering method involves bombardment by argon ions to dislodge silicon atoms from a target and deposit them onto the film. It is slower than evaporation and is generally regarded as not cost-effective in comparison to electron beam evaporation and chemical plasma deposition.

Chemical plasma deposition is quite different from the other methods. It uses a silicon-containing gas such as tetramethyldisiloxane or hexamethyldisiloxane, plus oxygen and helium. Power is applied to create a chemical plasma, activating the oxidation of the gas, and creating the SiO_x coating on the film surface. Because minimal heat is developed, plasma deposition can be used with heat sensitive materials such as LDPE and oriented PP. The degree of vacuum required is less than that for evaporation. Plasma deposition also tends to produce a thinner and clearer coating, with less yellowing. In addition, it is not a line-of-sight process like vacuum metallizing, evaporation, and sputtering, so it can coat in a three dimensional space, and can be applied to bottles and jars as well as to film. PET film 12 microns (0.5 mil) thick, coated by this process, exhibited oxygen barrier improvement from 115 $cm^3/(m^2$ d) (7.4 $cm^3/100$ in^2 d) to 1.1-2.0 $cm^3/(m^2$ d) (0.07-0.13 $cm^3/100$ in^2 d), depending on coating parameters. The barrier for oriented nylon improved from 40 $cm^3/(m^2$ d) (2.6 $cm^3/100$ in^2 d) to 0.5-7.0 $cm^3/(m^2$ d) (0.03-0.45 $cm^3/100$ in^2 d). Water vapor transmission rates improved as well [15].

An important additional advantage of SiO_x coating compared to vacuum metallizing, beyond its use in producing transparent and microwavable packaging, is that it does not interfere with metal detection equipment. These coatings are also reported not to change the recyclability of the base materials.

3.4.1.3 Aluminum Oxide Coatings [13,14]

Aluminum oxides can be used in place of silicon oxides, and combinations of SiO and MgO have also been used. These also produce transparent films with enhanced barrier properties. An evaporation technique using electron beams deposits the aluminum oxide on the base resin. Improvement in barrier performance is reported to be only moderate for oxygen, but the water vapor barrier increases to 3 to 10 times that of the base plastic substrate.

3.4.2 Laminations

Laminations are produced by taking individual webs of material and joining them into a composite. The individual layers are typically plastic, paper, or aluminum foil, and a large variety of combinations exist. There are a number of different methods for doing the combining, as well.

Adhesive lamination depends on the use of adhesives to "glue" the substrates (individual webs) together into a single structure. The adhesives themselves can be solvent or water-based, with water-based adhesives predominating. In the "wet-bonding" process, the adhesive is applied to one of the substrates, the substrates joined, and the lamination dried in an oven to evaporate the adhesive and complete the bonding process (Fig. 3.8). For this to work, at least one of the substrates must be porous to allow the water or organic solvent to escape.

In "dry-bonding," the adhesive is applied to a single substrate, dried, and then the second substrate is brought into contact with the first and adhered to the still tacky adhesive (Fig. 3.9). In hot-melt laminating (Fig. 3.10), a 100% solids, hot-melt adhesive is heated until it becomes liquid, and applied to the substrates, joining them as it cools.

In extrusion laminating, a melted polymer provides adhesion between the substrates as it cools, much like in hot-melt laminating. In extrusion laminating, however, a polymer such as polyethylene is melted in an extruder, while in hot-melt laminating, a hot-melt adhesive, a mixture of polymers and waxes which is easier to melt, is utilized (Fig. 3.11). In thermal laminating, heat activates a heat-seal coating on one web, which then joins with the second web (Fig. 3.12).

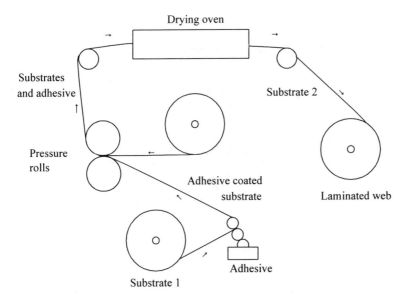

Figure 3.8 Lamination using wet bonding

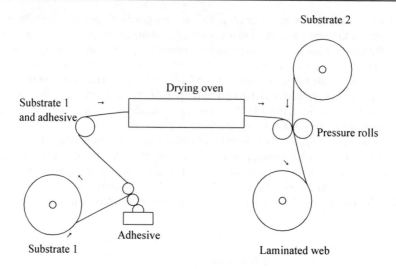

Figure 3.9 Lamination using dry bonding

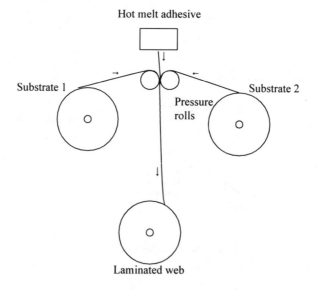

Figure 3.10 Hot-melt laminating

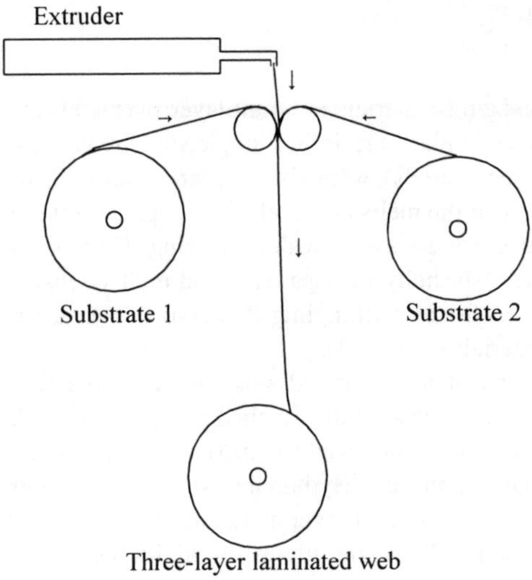

Extruder

Substrate 1

Substrate 2

Three-layer laminated web

Figure 3.11 Extrusion laminating

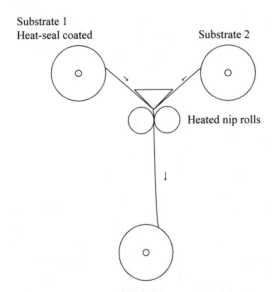

Substrate 1
Heat-seal coated

Substrate 2

Heated nip rolls

Laminated web

Figure 3.12 Thermal laminating

3.4.3 Coextrusion [4-7]

Coextruded flexible packaging materials are multilayer plastic sheet or film constructions produced from more than one plastic resin in a single step, by either the cast or blown film process. In coextrusion, the materials never exist as separate webs. Each type of resin is melted in an individual extruder, and the melts are carefully brought together in the die in a manner that keeps them in homogeneous layers, without mixing. Careful attention is required to rheology of the polymers, especially flow patterns and melt viscosity, in the design of the coextrusion to prevent layers from intermingling. Processing downstream of the die is identical to that of single layer materials (Fig. 3.13).

For many combinations of materials, adhesion between the two desired polymers is insufficient to result in good integrity of the finished product, and the layers tend to separate under stress. In this case, it is necessary to add a third component, known as a tie-layer, which is essentially a thermoplastic adhesive. By bonding strongly with both components, the tie layer maintains the integrity of the multilayer material. Tie layers must also must be melted in a dedicated extruder and fed into the die as an additional distinct layer. Thus, many coextruded materials have three-layer structures, one of each of the desired plastics, and a third

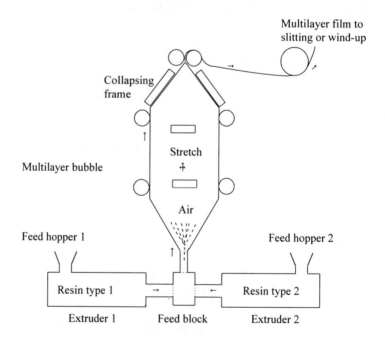

Figure 3.13   Blown film coextrusion

for the tie layer to provide adhesion between those two components. Structures with more than four layers are also common. For example, if EVOH is included as a barrier layer, it must be sandwiched between other polymers for moisture protection. With the addition of tie layers, this means at least a five-layer structure. Each new resin type in the structure requires a dedicated extruder, although the melted resin from a single extruder may be the source of two or more layers in the coextruded structure.

Coextruded films and sheet are used in a variety of packaging applications, including pallet stretch wrap with different cling characteristics on the inner and outer surfaces, snack food bags with brown-pigmented polyethylene in an inner layer to provide light protection and white-pigmented polyethylene on the inner and outer layers for a more desirable appearance, and polypropylene films with an ethylene vinyl acetate or low density polyethylene inside layer for improved heat-seal.

3.5 Stretch and Shrink Wrap

In a wrap, the simplest form of flexible plastic package, a flat piece of plastic film is folded or wound around the packaged item. Wraps are used to an extent for retail packages, but stretch wrap for pallet loads of goods to stabilize them during distribution is by far the largest market for this package form.

3.5.1 Stretch Wrap [16]

Stretch wrap exerts its holding power on its contents because the film is operating within the elastic portion of its stress-strain curve. The film has had stress exerted on it, and its molecules are now attempting to return to their original conformation. As long as its elastic limit is not exceeded, the greater the tension applied (and the greater the amount of stress), the greater the force exerted on the load by the film, and the greater the holding power of the film on the contents. Because the extent of stretching done during pallet wrapping is limited by the potential to crush the load and other factors, it is common to pre-stretch the film before it is wound on the load. Elongations of up to 250% can be achieved in this manner. Pre-stretching also minimizes necking-in (the tendency of the film width to decrease when the film is stretched).

Typically, the end of the stretch film is attached to the rest of the load simply by self-cling, which can be increased by adding tackifiers such as polyisobutylene to the base resin. A very smooth outer surface of the film also improves cling. It is important for a stretch film to have very high elongation, low neck-in, high tensile strength in the machine direction, high elasticity, good puncture resistance, and resistance to tear propagation, as well. High creep and fatigue resistance are very important, as they relate directly to the ability of the film to sustain its holding power over time. The holding force exerted by stretch wrap decreases with time because of stress relaxation in the plastic. Increasing temperatures increase the rate of stress relaxation.

Stretch films are most often LLDPE, modifed with various additive packages to yield the desired combination of properties. Coextrusions are frequently used to provide features such as a side with much stronger self-cling than the other side, so that the film sticks to itself as it is wound on the pallet, but does not adhere to adjacent pallet loads. EVA copolymers and PVC are also used for pallet stretch wrap, with the second also finding many applications in packaging meat, poultry, and fruits and vegetables.

3.5.2 Shrink Wrap [17]

Shrink films, like stretch films, are designed to provide a tight wrap around the packaged object through the tendency of the film to try to return to a smaller dimension. In shrink film the film starts out with larger dimensions than the packaged object, and then is exposed to heat, causing it to shrink. The restraint of the packaged object on the film's tendency to shrink provides the force between the film and the product.

In shrink film, a polymer film is oriented at an elevated temperature Then the orientation is "frozen" by quick cooling. When the film is again heated to roughly the orientation temperature, the oriented molecules will return to a more nearly random coil conformation, causing the film to shrink, because of the molecular "memory" of the polymer molecules.

To shrink wrap an object, a pouch is sealed loosely around it. Pouch and contents are then sent through a shrink tunnel, where the package is exposed to heat. If the temperature, residence time, and size of the package and product are properly chosen, the result is a tightly wrapped package.

Shrink wraps tend to loosen somewhat over time as a consequence of creep in the film. As is the case for stretch wrap, the loss of holding power is accelerated at higher temperatures.

A variety of films are used for shrink wrapping, including PE, PP, and PVC. Most are used for packaging consumer products. Some shrink wrap is used for pallets, but that market has been taken over by stretch film, with its lower energy requirements. Shrink film is also used in window insulation. The film is placed on the inside of a window and shrunk by using a hair dryer as the heat source.

3.6   Pouches

The next step in complexity from wraps is the general category of bags, sacks, envelopes, and pouches. All of these are made by folding and sealing the plastic film after it is blown or cast. These categories are not clearly distinguishable from each other in many cases, and can be generically described as "pouches."

Pouches all have at least one seam, most frequently formed by heat-sealing, although other methods, such as adhesives, may be used. Most pouches can be classified into one of three groups according to their geometry: pillow pouches, three-side seal pouches, and four side-seal pouches. Another important categorization separates those that can stand upright without support (stand-up pouches) and those that cannot.

A wide variety of items are packaged in plastic pouches, including products as diverse as books, pet food, lettuce, milk, shampoo, computer diskettes, and toys.

3.6.1 Pillow Pouches

Pillow pouches have back, top, and bottom seams (Fig. 3.14), which give them their characteristic "pillow" shape. Potato chip bags are a common example of pillow pouches. The back seam of the pillow pouch is usually formed as a "fin" seal. The two inner layers of the material are brought together and sealed, and the seal protrudes from the back of the finished pouch in such designs. Another option is a "lap" seal in which the inner layer of one side is

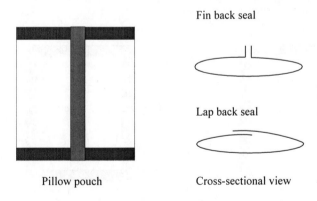

Fin back seal

Lap back seal

Pillow pouch   Cross-sectional view

Figure 3.14   Pillow pouches and fin and lap seals

sealed to the outer layer of the other side of the pouch, producing a flat seal. The fin seal is more common because it produces a stronger seal and requires only the inner layer of the material to be heat-sealable. The lap seal uses slightly less material, but is weaker and requires both the inner and outer layers of the pouch to be heat-sealable. Pillow pouches can be produced with or without gussets. The top and bottom seals are made between inner layers of the pouch material.

3.6.2 Three-Side Seal Pouches

Three-side seal pouches are produced with two side seams and a top seam (Fig. 3.15). The bottom of the pouch is produced by folding the material horizontally, and all seals are made between inner layers of the pouch material. These pouches also can be made with or without gussets. When gusseted, they can stand erect on the shelf. A variation on three-side seal pouches is to seal the side with the fold in addition to the other sides, thus essentially transforming it into a four-side seal pouch.

3.6.3 Four-Side Seal Pouches

Four-side seal pouches are usually produced from two webs (rolls) of material, sealed at the top, bottom, and both sides (Fig. 3.15). In contrast to the pouches described earlier, it is possible for the front and back of the four-side seal pouch to be made from two different plastic resins, as long as they can be sealed to each other. Four-side seal pouches can be made in a wide variety of shapes, such as hearts and ovals.

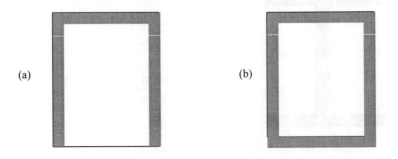

(a)   (b)

Figure 3.15 Pouches, (a) three-side seal, (b) four-side seal

3.6.4   Stand-Up Pouches [18]

Stand-up pouches can support themselves in an upright position on the shelf, at least when filled with product. They contain gussets which expand when the product is filled into the pouch, providing a relatively flat base for the pouch to stand on. They are widely used for both dry products and liquids, and their use is increasing as they replace more rigid structures such as cans and bottles. Stand-up pouches use less material than more rigid structures, and are thus generally more economical.

The first commercial use of standup pouches occurred approximately 35 years ago. The usual design incorporates a horizontal bottom gusset, most often made from a separate piece of flexible packaging material. In other applications, a single piece of film is used, folded into a W shape to produce the base. An alternative design is to use side folds or gussets instead of bottom gussets, with an overlapping flat sheet base. These pouches tend not to stand up as securely as those with a bottom gusset, making them more useful for dry products than for liquids. In both cases, the presence of sufficient product weight to spread out the gussets is necessary to make the pouch stand up.

Alternative designs form a rectangular bottom shape, or use side gussets that permit the bottom of the package to fold into a rectangular shape. These pouches can, if erected, stand up even without containing product.

Plastic stand-up pouches remained in quite limited use until the environmental pressures of the mid-to-late 1990s, when their ability to package products with considerably less material than more conventional packaging opened new opportunities. Improvements in technology to permit use of form-fill-seal equipment, and to incorporate convenient dispensing and reclosure features, such as spouts and zippers, into the pouches added to their appeal. On the other hand, pouches cannot support a load, and consequently require stronger distribution packaging. Further, filling equipment for pouches is considerably slower than for rigid containers, and pouches are also more difficult to handle. Therefore pouches often do not yield significant cost savings for the product manufacturer, despite the use of less material.

Stand-up pouches have had the greatest commercial success in Japan, particularly retort pouches (Section 1.6) for liquid food products. They are also widely used in India, especially for dry rice and automotive fluids such as motor oil. In Latin America stand-up pouches are used for a number of food products, as well as fabric softeners and detergents. In Europe, the major markets are laundry products such as detergent, personal care products such as lotion and shampoo, windshield-washer fluid, and latex paint. Markets in Canada include shampoo, liquid household cleansers, and motor oil.

The pouches have been less successful in the U.S. The first large volume product in stand-up pouches was Capri Sun fruit-flavored drinks (trademark of Capri Sun, Inc.). Stand-up pouches are also used for other products, including snack foods and detergents, to a limited extent. One of the factors limiting their use in North America was unavailability of pouches and machinery from a domestic source, thus increasing their cost. This is now changing, so use may increase in the future.

3.6.5   Forming Pouches

The most common pouches are pillow pouches and three-side seal pouches. Most frequently, products are packaged in form-fill-seal machines, in which pre-printed roll stock is formed into a package and the package is filled and sealed with product, all in a continuous operation within one piece of equipment. Pouches are usually cut apart within the machine, as well.

Form-fill-seal machines run in two configurations, vertical and horizontal, determined by the direction of travel of the package through the machine. The pouches are oriented vertically in a vertical form-fill-seal machine, and either vertically or horizontally in a horizontal form-fill-seal machine (Figure 3.16 and 3.17). A wide variety of pouch types can be made on either type of equipment, utilizing a single web or two webs. Pouches may be cut and sealed simultaneously, or may be sealed first and cut at a subsequent station in the machine [19,20].

Packaging operations may also use preformed pouches, rather than form-fill-seal equipment. In this case, the preprinted pouch is supplied ready to be filled with product and then sealed at the top. Filling and sealing, in this case, are generally done on two separate pieces of equipment.

There are advantages and disadvantages to both form-fill-seal and preformed pouches. For relatively large operations, with relatively easy-to-seal pouches, form-fill-seal operations are

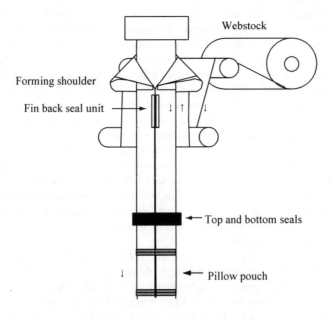

Figure 3.16   Vertical form-fill-seal machine

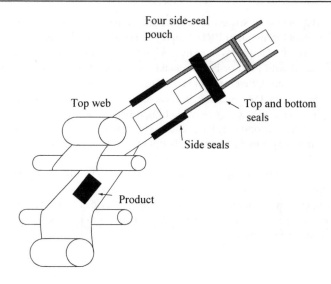

Four side-seal pouch

Top web

Top and bottom seals

Side seals

Product

Figure 3.17 Horizontal form-fill-seal machine using two webs

usually the most economical. The use of preformed pouches requires less capital investment since the equipment is simpler and less expensive. Quality control demands are higher in form-fill-seal operations, since the packager is responsible for the quality of all the seals. Thus, for hard-to-seal materials, there are advantages to using preformed pouches and shifting some of that responsibility to the pouch supplier. Most moderate-to-high volume packaging pouch operations use form-fill-seal technology. However, most operations using retort pouches or stand-up pouches employ preformed pouches.

3.6.6 Bulk and Heavy-Duty Bags [21-22]

Bulk bags and heavy duty bags are designed for packaging large quantities of product, up to 5,000 kg (11,000 lbs) in some cases. The bags have high tensile strength and can be suitable for containing either liquid or solid products. The most common material used, especially in the larger sizes, is woven PP fabric, though HDPE, PVC, and polyester fabric are also used. Some bags, especially in smaller sizes, are made of film rather than fabric, usually LLDPE or HDPE. Other bags include a plastic layer in a mostly paper construction, either as a film or as a coating on the paper. HDPE, LDPE, PVDC, PP, and combinations are all used in these applications.

Bags made of woven fabric and containing liquid products generally contain a PE film liner. Aluminum foil or PVDC copolymers can be used when a better barrier is required. The liner may be a single layer, or have a multi-material construction. Alternatively, the bags may be coated with PVC or latex to make them waterproof. Liners can also be used on bags for solid products. Often the liner is designed to be disposable to facilitate reuse of the bag itself.

Seams on the bag may be heat-sealed or sewn, depending on the material and the application. The bag may be made from tubular material, and thus require sealing only at the top and bottom, or it may have side and bottom seams. Some bags are made like three-side seal pouches, with a fold forming the bottom of the bag. Both filling and discharge openings may be simply tied shut rather than sealed.

Large size bulk bags are generally designed to be filled from the top and emptied from the bottom. Spouts or other dispensing devices may be included, or the bags may simply be cut to open. They usually incorporate special devices, such as loops of fabric, to facilitate handling. For bags stored outside, it is important to incorporate an ultraviolet light stabilizer into the resin formulation.

3.7 Heat Sealing [23]

Heat sealing generally relies on a combination of heat and pressure to fuse two layers of thermoplastic material together. A heat seal is strongest when the interface between the materials has disappeared, but the materials being fused have not thinned or degraded. Many types of heat sealing are used in packaging. While some of these methods are applicable to rigid rather than flexible packaging, they are all discussed here.

3.7.1 Bar or Thermal Sealing

Bar sealing, or thermal sealing (Fig. 3.18), involves placing the materials to be sealed between heated bars, which press them together. Heat is conducted through the materials to the interface, and the materials are fused. When sufficient time has elapsed to form the seal, the bars release and the hot material is moved out of the seal area. The seal does not have its full strength at this point, but must be sufficiently strong to hold together (hot tack). Full strength develops as the seal area cools to the ambient temperature. If too little heat, residence time, or pressure is applied, an adequate seal will not be formed. If too much heat, time, or pressure, is applied, excessive flow occurs in the heat seal layers, weakening the material. The edges of the sealing bars are generally rounded to avoid puncturing the packaging material. The film-contact surface of one of the bars may be resilient to aid in achieving uniform pressure on the seal area.

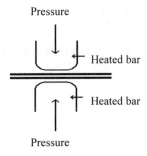

Figure 3.18 Thermal bar sealing

A variation on bar sealing is to use one heated bar and one bar which is not heated. In this case, all the heat conduction occurs in one direction only. Another variation is to pass the materials to be sealed through heated rollers. A disadvantage of this method is the very short contact time, necessitating either preheating, slow travel through the rollers, or both.

Bar sealing is the most commonly used method of heat-sealing packaging materials, especially for form-fill-seal operations. It can also be used for sealing lids on cups and trays, with the upper bar shaped to match the rim of the container.

3.7.2 Impulse Sealing

In impulse sealing, the material to be sealed is placed between two jaws, one or both of which contain a nichrome wire ribbon (Fig. 3.19). The jaws apply pressure to the materials, and an impulse of electrical current (less than one second) is passed through the wires, causing them to heat. After the pulse of current, the materials are retained between the jaws for a predetermined dwell time, allowing cooling to occur under pressure. This method enables sealing of materials with insufficient hot tack (adhesion to each other when hot), or materials that are too weak to permit unsupported travel at the sealing temperature. In some impulse sealers, water cooling of the sealing jaws is provided. Shaped impulse seals can be used for sealing lids on cups and trays, just as is done for bar sealing.

Impulse sealing generally gives a better looking seal than bar sealing, since the seal is narrower. For the same reason, impulse seals are often not as strong as bar seals. An additional problem with impulse sealing is relatively high maintenance requirements. The nichrome wires tend to burn out, and the fluoropolymer tape used to keep the plastic from sticking to the bars tends to need frequent replacement. However, impulse sealing remains a commonly used method of sealing packaging materials.

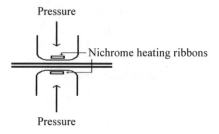

Figure 3.19   Impulse sealing

3.7.3   Band Sealing

Band sealing is another method for providing a cooling phase under pressure as well as a heating phase. Here, the materials to be sealed are carried along by two moving bands past a heating station and a cooling station (Fig. 3.20). This system allows for very high speed sealing, but the sealed materials often wrinkle in the process. Band sealing is often used for sealing pouches that have been filled with product.

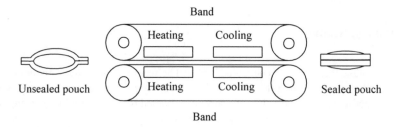

Figure 3.20   Band sealing

3.7.4   Hot Wire or Hot Knife Sealing

A very fast sealing method is to use a hot wire or a hot knife to weld together and cut apart plastic films at the same time. The wire or knife travels through the two layers of film, melting them and cutting them apart from the rolls of webstock. The seal produced is very tiny and often nearly invisible. This method cannot be used where a hermetic seal (a barrier to microorganisms) is required, but is excellent for relatively undemanding packaging produced at very high speeds. LDPE is the plastic most commonly used in such applications.

3.7.5 Ultrasonic Sealing and Friction Sealing

Heat for sealing can be generated by friction instead of by conduction from a hot surface. This is the principle behind ultrasonic and friction sealing. Ultrasonic sealing involves hammering or rubbing together two surfaces at a high frequency, generating heat which then produces a weld. This method is especially useful when the materials to be sealed are too thick for efficient heat transfer, or when they are highly oriented and would lose too much orientation if exposed to heat for a significant period of time. Friction sealing, or spin welding, is the same principle applied to containers. The top and bottom half of cylindrical containers are sealed to each other by rotating one rapidly while the other is held in place (see Section 4.12). Friction again produces heat and welds the parts together. This method can also be used for sealing caps on bottles, and sometimes to seal containers with shapes other than round by employing oscillation rather than rotation.

3.7.6 Hot Gas and Contact Sealing

Another method that eliminates conduction of heat through the materials to be sealed is to directly heat the sealing surfaces before bringing them together. In hot gas sealing, a gas flame or hot air is used to heat the surfaces and melt them. The surfaces are then pressed together between chilled jaws to complete the sealing process. Contact sealing uses a heated plate instead of the hot gas to melt the surfaces to be sealed.

3.7.7 Radiant Sealing

For materials subject to excessive distortion under pressure, radiant sealing can be used. In this method, a radiant heat source melts the material, and no pressure is applied. This method is often used for sealing uncoated PET, spunbonded materials for medical device packaging (such as Tyvek from DuPont, Inc.), and oriented films. A special advantage of radiant heating is that shaped seals can be easily produced by designing a radiant heater element with the desired shape.

3.7.8 Dielectric, Magnetic, and Induction Sealing

Dielectric, magnetic, and induction sealing are another group of methods which have in common the effort to generate heat rapidly in the seal area, without depending very much on conduction.

In dielectric heating, the material to be sealed is exposed to a high frequency electrical field. If the material is polar, the attempt of the molecules to align appropriately with the rapidly oscillating field generates heat, and consequently a seal. This method will not work with nonpolar polymers such as polyolefins, but is generally the method of choice for heavy PVC materials, especially for textured PVC.

Magnetic sealing uses an oscillating magnetic field instead of an electrical field. If the material to be sealed contains magnetic iron compounds, they attempt to line up with the magnetic field, again producing heat. Since packaging materials do not, in general, contain iron compounds, these materials are usually supplied in the form of a gasket or a coating. The magnetic field melts the gasket, which in turn melts the sealing surfaces. This method is sometimes used for sealing cap liners or lids.

Induction sealing is widely used for producing tamper-indicating inner seals on plastic bottles and jars. In induction sealing, an aluminum foil-containing structure is placed in an alternating magnetic field. The field induces an electrical current in the foil. The current heats the foil, causing adjacent structures to be heated, and a seal to form. The original inner seal structure for induction sealing, still in use today, consisted of a heat seal coated foil attached with wax to a paperboard backing, which was glued into the closure (see Fig. 3.21). When the foil was heated, the heat seal coating was activated and sealed to the bottle rim. The wax melted and was absorbed by the paperboard backing, which then freed the foil from the cap. In addition to a tamper-indicating feature, this type of seal provides an excellent barrier to protect the product against gain or loss of moisture, oxygen, and other components.

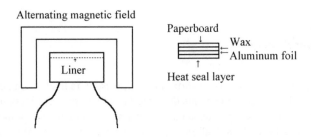

Figure 3.21   Induction sealing, showing original cap liner composition

3.8 Other Sealing Methods

Not all sealing methods depend directly or indirectly on heat. Adhesives can, of course, be used. Another option is solvent sealing. In this method, a solvent is applied to the surfaces to be joined together. With time, the solvent dissolves some of the surface of the materials, and the surfaces are then pressed together. The wetted, viscous interfaces that have thus been formed blend together. The material is held under pressure long enough for sufficient solvent to diffuse into the bulk of the material, away from the interface, that the resultant seal has enough tack to hold together. The seal continues to increase in strength as the residual solvent migrates away from the seal area, eventually diffusing into the air. This method avoids the application of any heat in achieving the seal, but the cost in most cases is exposure of both the packaging material and the environment to organic solvents. Excessive solvent may weaken the packaging material, and there are regulatory, employee health and safety, and environmental concerns related to solvent release, recapture, and disposal. In a few systems, such as those involving polyvinyl alcohol, water can be used as the solvent, eliminating these problems.

3.9 Evaluation of Seals in Flexible Packaging Materials [15]

In most packaging applications, the goal is to achieve a seal at least as strong as the rest of the package. A common test is to subject the seal area to enough tension to cause failure and observe the failure location. If the failure occurs in the seal area, the seal is the weak point and is not acceptable. If the failure occurs in the bulk material, then the seal is acceptable. This failure will usually be near the seal, because of some thinning of the material adjacent to the seal during the sealing operation. A major exception to this rule is the case where a peelable seal is desired, generally for ease of opening the package. In this case, it is desirable to have the seal less strong than the package material and for failure to occur at the seal interface by separation of the sealed layers from each other.

Polarized light can be used to examine heat seals in transparent materials. The birefringence patterns obtained are a visual guide to the consistency of the seal throughout the seal area, and can readily show gaps and stress areas.

When a hermetic seal is needed, it is imperative to avoid wrinkles. The presence of small wrinkles often does not significantly affect the mechanical strength of seals, but does affect their ability to keep out microorganisms. The best way to avoid wrinkles in a seal is to keep bidirectional tension on the material as it is being sealed. Contamination in the seal area can similarly provide entry to microorganisms.

3.10 Advantages and Disadvantages of Flexible Packaging

The major advantage of flexible packaging is economy. It, in general, makes very efficient use of both materials and of space. The ratio of delivered product to package material is large. The packages tend to be efficient in their use of cube, meaning distribution packaging can be smaller. Storage of unfilled packages occupies minimal space, especially if they are stored as webstock. The packages are generally quick and easy to form, which adds up to a low-cost package.

The primary disadvantages of flexible packaging are its lack of strength and relative inconvenience for the user. Flexible packaging has virtually no ability to support a load, thus placing greater demands on distribution packaging. More importantly to consumers, flexible packages tend to be difficult to open and often impossible to reseal effectively.

The issue of consumer convenience has been addressed by a number of relatively new developments. One of the oldest opening and resealing systems, and still one of the most convenient, is a zipper closure. Other developments include easy-peel seals, some of which are also capable of reclosure, and the incorporation of spouts, which may incorporate a regular closure (cap) as is used on bottles. Of course, the incorporation of such devices adds substantially to the package cost, but they can have a significant impact in increasing consumer acceptance, especially in markets where flexible packages compete against rigid containers.

An important way to compensate for the lack of rigidity and load-bearing capacity of flexible packaging is to combine pouches with paperboard cartons or with corrugated fiberboard boxes in bag-in-box packages [24,25]. These packages are used both for dry products, such as breakfast cereal, and liquid products, such as inexpensive wines. For oxygen-sensitive products like wine, these packages have a significant advantage. As the product is dispensed, the bag collapses, limiting the ingress of oxygen and thus prolonging the shelf life of the product.

Overall, the advantages of flexible packaging frequently outweigh the disadvantages, resulting in sustained rapid growth in this segment of the packaging industry.

References

1. *Modern Plastics Encyclopedia Handbook* (1994) New York: McGraw-Hill
2. Brody, A.L. and Marsh, K.S. (1997) In *The Wiley Encyclopedia of Packaging Technology, 2nd ed.*, A.L. Brody and K.S. Marsh (Eds.), New York: John Wiley & Sons, pp. 370-378
3. Berins, M.L. (Ed.) (1991) *Plastics Engineering Handbook of the Society of the Plastics Industry*, 5th ed., New York: Chapman & Hall
4. Bongaerts, H. (1988) In *Plastics Extrusion Technology*, F. Hensen (Ed.), Munich: Hanser Pub., pp. 143-202
5. Hensen, F. (1988) In *Plastics Extrusion Technology*, F. Hensen, Ed., Munich: Hanser Pub., pp. 257-283

6. Winkler, G. (1988) In *Plastics Extrusion Technology,* F. Hensen (Ed.), Munich: Hanser Pub., pp. 95-124

7. Hessenbruch, R. (1988) In *Plastics Extrusion Technology,* F. Hensen (Ed.), Munich: Hanser Pub., pp. 125-142

8. Hanlon, J.F. (1992) *Handbook of Package Engineering,* 2nd ed., Lancaster, Pa: Technomic Pub. Co, Inc.

9. Prince, P.E. (1986) In *The Wiley Encyclopedia of Packaging Technology,* M. Bakker (Ed.), New York: John Wiley & Sons, pp. 443-446

10. Bakish, R. (1997) In *The Wiley Encyclopedia of Packaging Technology, 2nd ed.,* A.L. Brody and K.S. Marsh (Eds.), New York: John Wiley & Sons, pp. 629-638

11. *Mylar Polyester Film for Packaging, Summary of Properties, Type MMC Metallized Coated MYLAR Polyester Film* (1980) Wilmington, Delaware: DuPont Company

12. Toensmeier, P.A. (Feb. 1995) Modern Plastics, pp. 17-18

13. Brody, A.L. (Feb. 1994) Packaging Technology & Engineering, pp. 44-47

14. Finson, E. and Hill, R.J. (June 1995) Packaging Technology & Engineering, pp. 36-43

15. Hill, R.J. (1997) In *The Wiley Encyclopedia of Packaging Technology, 2nd ed.,* A.L. Brody and K.S. Marsh (Eds.), New York: John Wiley & Sons, pp. 445-448

16. Le Caire, R. (1986) In *The Wiley Encyclopedia of Packaging Technology,* M. Bakker (Ed.), New York: John Wiley & Sons, pp. 338-341

17. Jolley, C.R. and Wofford, G.D. (1997) In *The Wiley Encyclopedia of Packaging Technology, 2nd ed.,* A.L. Brody and K.S. Marsh (Eds.), New York: John Wiley & Sons, pp. 431-434

18. Greely, M.J. (1997) In *The Wiley Encyclopedia of Packaging Technology, 2nd ed.,* A.L. Brody and K.S. Marsh (Eds.), New York: John Wiley & Sons, pp. 852-856

19. Brody, A.L. and Marsh, K.S. (1997) In *The Wiley Encyclopedia of Packaging Technology, 2nd ed.,* A.L. Brody and K.S. Marsh (Eds.), New York: John Wiley & Sons, pp. 465-468

20. Brody, A.L. and Marsh, K.S. (1997) In *The Wiley Encyclopedia of Packaging Technology, 2nd ed.,* A.L. Brody and K.S. Marsh (Eds.), New York: John Wiley & Sons, pp. 468-470

21. Twede, D. (1997) In *The Wiley Encyclopedia of Packaging Technology, 2nd ed.,* A.L. Brody and K.S. Marsh (Eds.), New York: John Wiley & Sons, pp. 51-54

22. Doar, L.H. Jr. (1997) In *The Wiley Encyclopedia of Packaging Technology, 2nd ed.,* A.L. Brody and K.S. Marsh (Eds.), New York: John Wiley & Sons, pp. 60-61

23. Brody, A.L. and Marsh, K.S. (1997) In *The Wiley Encyclopedia of Packaging Technology, 2nd ed.,* A.L. Brody and K.S. Marsh (Eds.), New York: John Wiley & Sons, pp. 823-827

24. Brody, A.L. and Marsh, K.S. (1997) In *The Wiley Encyclopedia of Packaging Technology, 2nd ed.,* A.L. Brody and K.S. Marsh (Eds.), New York: John Wiley & Sons, pp. 46-48

25. Arch, J. (1997) In *The Wiley Encyclopedia of Packaging Technology, 2nd ed.,* A.L. Brody and K.S. Marsh (Eds.), New York: John Wiley & Sons, pp. 48-51

4   Thermoformed Packages

4.1   Introduction

Thermoforming is the process of using heat to soften a piece of plastic sheet (or film), and then pressure to shape the plastic as desired. Most often a mold is used, with a vacuum drawn on one side, and differential air pressure to provide the molding force. Thermoforming can be used to make a variety of package shapes, including cups, trays, egg cartons, and blister packages in a relatively inexpensive way. Molds are generally less costly than those for injection or blow molded packaging components, as they are simpler and typically not subject to high pressures, and production speeds can be very high. It is estimated that in the U.S. more than 75% of thermoforming applications are packaging [1]. Polystyrene (PS) is the resin most often used for thermoforming, especially high impact polystyrene. PVC is very widely used also, especially for blister packaging.

Packaging is generally thermoformed from webstock, rather than from cut sheets, although these are used for some heavy industrial applications, such as pallets and large bins. Placing the thermoforming operation in-line with the extruder eliminates the need to roll up the web.

Parts almost always require trimming, which can be done in the mold, at another station in the machine, or downstream. Scrap rates are often fairly high, but in most cases the scrap can be recycled to sheet manufacture.

Control of the temperature to which the sheet is heated is very important. If the sheet is not hot enough, it will not form well, and may either rupture or fail to reach all parts of the mold. On the other hand, if it is too hot, it may flow excessively, resulting in thin areas and lack of expected orientation. Heating is done by radiation, convection, or a combination of the two. Conduction heating can also be used, but is uncommon. For packaging applications, infrared radiation is by far the most common heat source [1,2].

Cooling is accomplished by conduction to the mold and convection to the surrounding air. Parts must be cooled below their heat deflection temperature before they are removed from the mold. Molds are sometimes, but not always, equipped with cooling mechanisms.

4.2   Positive and Negative Thermoforming [1-3]

Two basic types of thermoforming are positive, or male mold forming, in which the shape of the mold is convex (Fig. 4.1), and negative, or female molding, in which the shape of the mold is concave (Fig. 4.2). The mold shapes one side of the package, with the other side exposed to air.

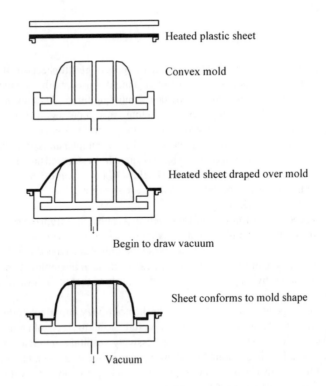

Figure 4.1   Positive thermoforming (convex, male mold)

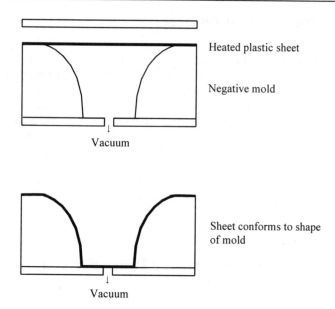

Heated plastic sheet

Negative mold

Vacuum

Sheet conforms to shape
of mold

Vacuum

Figure 4.2   Negative thermoforming (concave, female mold)

The choice of thermoforming method results in significant differences in the distribution of the plastic in the wall of the finished packaging component. The general principle is that, under pressure, hot plastic stretches. When the plastic touches the cooler mold, it stops stretching. The farther the plastic stretches, the thinner it gets.

4.2.1 Positive Thermoforming

In the basic thermoforming process with a convex mold, the heat-softened plastic is drawn down over the mold by drawing a vacuum through the mold, thus causing air pressure to force the plastic into contact with the mold. The hot plastic first touches the dome of the mold, and drapes down over the sides until it finally makes contact at the rim, completing the draw. Thus, the dome has been stretched the least and is the thickest, while the rim has been stretched the most and is the thinnest. The surface finish of the mold, including any desired markings, is imparted to the interior of the molded cavity. As the plastic cools, it shrinks and draws more tightly in contact with the mold. Therefore, the mold has to be designed with a large enough draft angle to permit removal of the part after cooling. Positive molding is preferred when product tolerances on the inside of the package are critical.

4.2.2 Negative Thermoforming

In the basic forming process with a concave mold, the heat-softened plastic is drawn down into the mold cavity by drawing a vacuum through the mold, causing air pressure to force the plastic into contact with the mold, and stretching it as it goes. The first contact between the plastic and the mold is at the rim of the cavity, so this will be the thickest area. The last contact is at the dome, particularly at the dome edges, so these will be the thinnest. The surface finish of the mold, including any desired markings, is imparted to the outside of the molded cavity. As the plastic cools and shrinks, it pulls away from the mold. Thus it is much easier to remove from the mold than a part of similar shape produced by positive forming, and draft angles can be nearer to 90°. Negative forming generally provides better material distribution and faster cooling than positive forming.

4.3   Vacuum (Straight) and Drape Forming

Another classification of thermoforming depends on how the plastic is moved into contact with the mold. In vacuum forming, also known as straight forming, the primary force causing the hot plastic to come into contact with the mold is a vacuum drawn through the mold, resulting in a difference in air pressure between the two sides of the plastic sheet, pushing the plastic against the mold.

In drape forming, a vacuum is also applied to remove trapped air. But first, the plastic is allowed to sag against the mold under the influence of gravity, which is considered the major forming force.

Most frequently, drape forming is used with positive molds, and straight vacuum forming with negative molds. Thus, many people use these sets of terms interchangeably. To add to the complexity, when multiple molds are used, there is often significant draping only around the edges, with primarily stretching action occurring in the centrally located molds, whether they are positive or negative.

4.4   Thermoforming Variations [1-3]

A number of variations of the basic types of thermoforming exist. They have all been devised to improve in some fashion on the basic processes by permitting deeper draws, more complex draws, better control over wall thickness, thermoforming of more difficult materials, among other objectives.

4.4.1 Pressure Forming

In vacuum forming, as described above, the forming pressure is the difference between the pressure (near zero) on one side of the sheet where the vacuum is being drawn, and that on the other side, which is at one atmosphere. Thus, the maximum forming pressure is one atmosphere (0.1 MPa, 14.7 psi). For some applications, this is simply not enough pressure to get a good draw. If the thermoforming equipment is designed to allow it, additional air pressure can be applied to the high pressure side of the sheet, up to 3.45 MPa (500 psi), although the typical range of pressure is 0.14 to 0.56 Mpa (20-80 psi) [1]. It should be noted that these are gauge pressures, so the actual forming pressure is the difference between the gauge pressure and the air pressure on the other side of the sheet. Vacuum is still used on the low pressure side to avoid problems caused by trapped air. Pressure forming is illustrated in Fig. 4.3.

Pressure forming requires the use of stronger, more substantial and expensive molds than vacuum forming. Equipment is also more complex and more costly. Negative molds are generally used.

Pressure forming is typically used for difficult-to-form materials such as PP, and for the production of highly detailed parts.

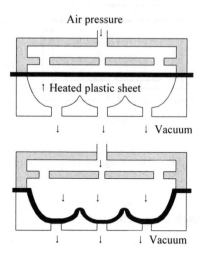

Figure 4.3   Pressure forming

4.4.2 Plug-Assist Thermoforming

Plug assist thermoforming (Fig. 4.4) uses a negative mold coupled with a mechanical plug to help to move the plastic into the mold cavity. The plug is typically 60-90% of the volume of the cavity [2]. It permits better material distribution in the finished package and deeper draws. The plug's size, geometry, and rate and depth of movement are all important variables, as is the relationship between the timing of the plug motion and the timing of the vacuum draw in the mold cavity. In general, the vacuum is not started until after the plug is fully engaged. In this case, the process resembles thermoforming with a positive mold followed by a negative mold, except that care must be taken to ensure that the plug does not cool the sheet too much for successful forming. Pressure forming and plug-assist forming are often combined to give plug-assist pressure forming, with the air pressure commonly supplied through the plug.

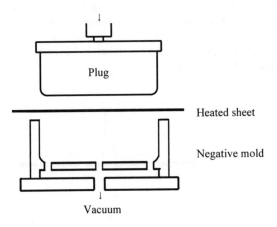

Figure 4.4   Plug assist thermoforming

4.4.3 Vacuum Snap-Back Thermoforming

Just as plug-assist forming can be considered positive forming followed by negative forming, vacuum snap-back forming (Fig. 4.5) can be thought of as negative forming followed by positive forming. In this thermoforming variation, there is a positive mold and a negative vacuum box. The heated plastic is first drawn partially into the vacuum box, using a low vacuum, as the positive mold is also lowered into the box. This vacuum is then stopped, and a high vacuum is drawn through the positive mold, causing the plastic to reverse its direction of travel, and "snap-back" around the positive mold. During this stage, positive air pressure can be applied through the vacuum box, if desired. The advantage of this method is that the prestretching of the bubble of plastic sheet gives better wall distribution in the finished product, allowing deeper draws while still permitting positive molds to be used. For additional stretching, bubble forming can be used before the initial draw into the vacuum box.

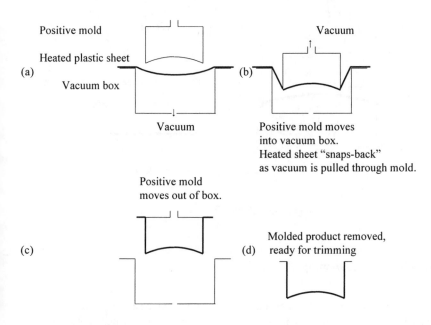

Figure 4.5   Vacuum snap-back thermoforming

4.4.4   Bubble or Billow Forming

The concept of pre-stretching the plastic sheet with air can also be used with negative molds. In this case, air pressure in the mold cavity causes the plastic to be stretched into a bubble above the cavity. Because there is no mold contact to cool the plastic, stretch is uniform. The stretched plastic is next drawn into the mold with vacuum, often coupled with plug-assist (Fig. 4.6). Additional air pressure may also be applied through the plug during the final forming. Again, deeper draws are possible and more uniform wall thickness results.

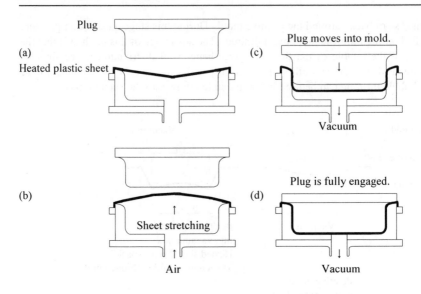

Figure 4.6 Bubble forming

4.4.5 Matched Mold Forming

The most complex draws, the best dimensional control, and the highest forming pressures can be achieved by avoiding reliance on air pressure altogether. In matched mold forming, mechanical pressure on both sides of the softened plastic sheet forms it between a negative and a positive mold. Vacuum is generally applied through the negative cavity (Fig. 4.7). The major application in packaging is for thermoforming foams, such as PS egg cartons.

4.5 Conventional and Solid-Phase Pressure Forming [2]

Thermoforming can also be classified for crystalline materials by the temperature at which the forming occurs. Conventional thermoforming refers to thermoforming of amorphous materials, or of crystalline materials at temperatures above their melting point (also referred to as melt phase thermoforming). It is also possible to thermoform crystalline plastics at temperatures slightly below their crystalline melting points, at temperatures 5-8% below the

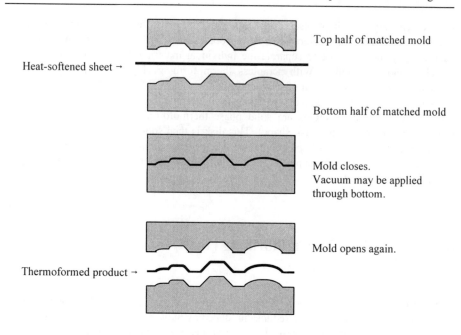

Top half of matched mold

Heat-softened sheet →

Bottom half of matched mold

Mold closes.
Vacuum may be applied
through bottom.

Mold opens again.

Thermoformed product →

Figure 4.7   Matched mold thermoforming

melt-phase forming range in a process called solid phase forming. Materials, as would be expected, are considerably more difficult to form under these conditions, and therefore require the use of the plug-assist and pressure forming techniques discussed earlier. The combination of plug-assist, pressure forming, and solid phase forming is known as solid phase pressure forming (SPPF).

SPPF can produce stiffer parts with less materials, and improves the sidewall strength of containers, while also increasing stress. Forming pressures of about 689 kPa are typically required, and plug design is more critical than in melt phase thermoforming [2].

4.6   Scrapless Thermoforming

SPPF allows the economical use of high oxygen barrier, coextruded materials in thermoforming. As mentioned earlier, thermoformed products, especially those with round shapes, typically have high scrap rates. For most products, this does not present a large problem, since the scrap material can be reground and reused. However, for coextruded material, it is not possible to separate the individual materials from each other, making recycling more difficult. With scrapless thermoforming (Fig. 4.8), it is possible to dramatically reduce the amount of scrap generated.

In this process, square segments are cut out of the rectangular sheet, generating little scrap. These relatively thick sections are solid phase thermoformed between matched molds, resulting in a much thinner disk shape. This disk is then solid phase plug-assist pressure formed into the final bowl-shaped container. The primary application is for microwavable containers for soups and similar products that require a multi-layer coextruded sheet containing an oxygen barrier plastic.

4.7   Twin-Sheet Thermoforming [1-3]

Another innovation involves thermoforming two sheets of plastic and welding them together to produce a hollow part, much like those produced by blow-molding. This process, known as twin-sheet thermoforming, has two basic variations. Sequential twin sheet thermoforming feeds one section of sheet, and thermoforms it into one side of the mold (usually the lower half), while a second section of sheet is being heated. Then the second sheet feeds in and gets formed into the second side (upper half) of the mold, at the same time the edges are welded together (Fig. 4.9). Machines usually are fed by cut sheets, rather than rollstock.

Simultaneous twin sheet thermoforming advances both sheets into the mold at the same time, forming one into one side of the mold and the other into the opposite side and welding the edges, simultaneously. A blow-pin is used to intruduce air pressure between the sheets to aid in the forming (Fig. 4.10). Machines are often fed by two rolls simultaneously.

In both methods, combinations of vacuum and air pressure are used to provide the forming force. A common application of twin sheet thermoforming in packaging is the production of plastic pallets.

Plastic hollow containers that can be opened and closed can be produced by conventional thermoforming using two adjacent cavities. The plastic material between the cavities becomes the hinge when the material is folded over. Some type of locking device (e.g., slot and tab) is incorporated to keep the folded-over container from opening back up.

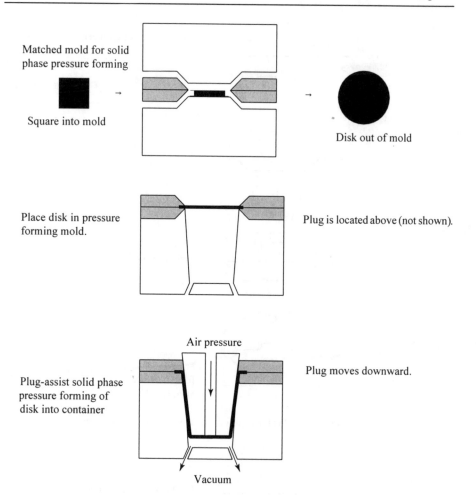

Matched mold for solid phase pressure forming

Square into mold

Disk out of mold

Place disk in pressure forming mold.

Plug is located above (not shown).

Air pressure

Plug-assist solid phase pressure forming of disk into container

Plug moves downward.

Vacuum

Figure 4.8   Scrapless thermoforming, utilizing solid phase pressure forming

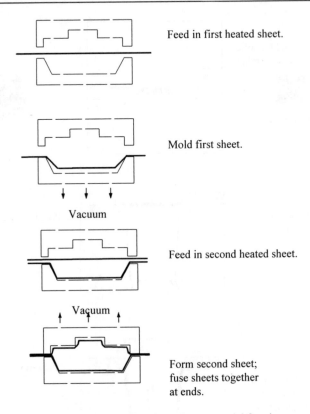

Feed in first heated sheet.

Mold first sheet.

Vacuum

Feed in second heated sheet.

Vacuum

Form second sheet;
fuse sheets together
at ends.

Figure 4.9 Twin-sheet thermoforming using sequential forming

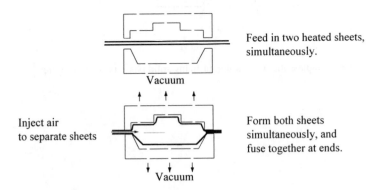

Feed in two heated sheets,
simultaneously.

Vacuum

Inject air
to separate sheets

Form both sheets
simultaneously, and
fuse together at ends.

Vacuum

Figure 4.10 Twin-sheet thermoforming using simultaneous forming

4.8 Melt-to-Mold Thermoforming

Melt-to-mold thermoforming is a process generally classified as thermoforming but similar to production of cast film. Here, the molds are located on a chilled chrome cylinder. The melted plastic is extruded through a slit-shaped die and deposited on the chrome cylinder, where it solidifies. The formed parts are then removed from the cylinder and trimmed. The result is a savings in energy since the sheet does not need to be reheated before it is thermoformed. In addition, thermal stress in the finished containers is significantly reduced.

4.9 Skin Packaging

A much older type of thermoforming is skin packaging. Skin packaging differs from conventional thermoforming because it does not require the use of molds, plus film, rather than sheet, is used. The product to be packaged is placed on a backing, usually coated paperboard, although plastic and uncoated corrugated board can also be used with certain films. The heat-softened film is lowered over the product and backing and drawn down tightly by applying vacuum. The film heat-seals to the backing, capturing the product and holding it tightly in place (Fig. 4.11). Skin packaging is also known as pneumatic sealing, and is used extensively for consumer products as an alternative to blister packaging

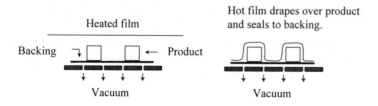

Figure 4.11 Skin packaging

Skin packaging, when used on precut backing, tends to be rather labor intensive. However, it is very economical for short production runs, where the type of product being packaged changes frequently. Highly automated continuous skin packaging systems are also available. Products that are packaged this way must be able to withstand brief contact with hot plastic without damage and also not adhere to the plastic.

The backing board is usually coated with a heat-seal material to secure good adhesion between the board and the film. However, ionomer films seal to uncoated paperboard, such as corrugated. The vacuum is typically drawn through the board, so coated boards or polymers must usually be perforated to get a good draw. Some systems evacuate the air through the gap between the two materials as the skin packaging is drawn down, but problems are often encountered in getting the nearly complete evacuation required for a tight package.

4.10 Thermoforming Molds [1-3]

Molds used for high-volume thermoforming applications are most frequently made from cast aluminum because of its excellent thermal conductivity, light weight, and reasonable cost. Water channels in the mold provide for the flow of cooling water, or a cored mold design can be used. Ideally, the difference between water inlet and outlet temperatures should be no more than 3° C. Alternative mold materials include wood, phenolics and other plastics, and steel. These molds are used for shorter volume runs, and generally do not include temperature control systems. Fans to blow air over the top of the mold can aid in cooling. Chrome-plated beryllium copper molds are sometimes used for especially long service life.

Molds need vacuum holes or slits for the removal of air between the mold and the plastic sheet. The mold surface is often sand-blasted to further prevent problems caused by trapped air, which can cause imperfections in the finished part. Sand-blasting is not generally used for HIPS, as it can make part removal difficult.

If undercuts are required to produce the package, a split mold may be needed, or a mold with a removable part that pulls out after thermoforming to release the package. In other cases, the package may be deformed enough to be released from the mold without requiring these features.

Generally, negative molds are less likely to be damaged, while positive molds are cheaper to produce. Draft angles on negative molds are typically at least 2 to 3°, while those on positive molds have to be 5 to 7°, in most cases.

4.11 Thermoform-Fill-Seal [4]

Just as products can be packaged in pouches in form-fill-seal equipment, they can be packaged in thermoformed containers in thermoform-fill-seal equipment (TFFS), typically in a horizontal configuration. Primary applications are for vacuum packaging of products such as meat and cheese, and for modified atmosphere packaging of products such as fresh vegetables. Liquid products such as yogurt can also be packaged in this fashion. Typically, the lower web is thermoformed, the product is filled into the cavities produced, a lidding web is indexed on, the air is evacuated, a modified atmosphere added if desired, and the lidding is sealed to the bottom web. In some cases, both the upper and lower webs are thermoformed. It is also possible for the upper web to be thermoformed and the bottom web not formed.

Either positive or negative forming can be used, as well as plug-assist and other modifications. The web emerges from the forming station with the cavity formed and ready for loading the product. The entire area of the top web not in contact with the product may be sealed to the bottom web, or a portion may be left unsealed to facilitate opening. One patented process injects steam into the sealing die to shrink the lower web tightly around the product, producing a combination of thermoforming and skin packaging. Evacuation of the air in the cavity is accomplished through the gap between the upper web and the lower web, after an outer box in the sealing die chamber has closed the package off from the surrounding air. Injection of a modified atmosphere into the package can also take place at this time (Fig. 4.12).

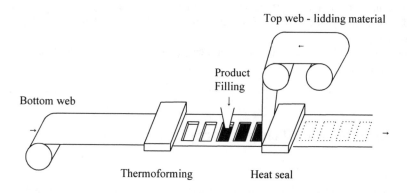

Figure 4.12 Thermoform-fill-seal packaging.

TFFS is commonly used for packaging medical devices, as well as food. In medical packaging, the lidding material is commonly a spun-bonded polyolefin, to allow for ethylene oxide sterilization.

4.12 Aseptic Thermoforming

In aseptic packaging, the need to sterilize the packaged product to ensure the destruction of harmful bacteria is eliminated. The product and the package are sterilized separately, and then the two are combined in a manner that maintains sterility. These techniques can be applied to thermoformed packages as well as to flexible packages and blow-molded containers.

A particularly ingenious application, although it is no longer sold, was the Conoffast process (a trademark of Continental Can), which used a coextruded bodystock and lidstock for the thermoformed package. As the body material entered the sterile environment, the coextruded sheet was heated and the top layer stripped off. This exposed a sterile inside surface. The container was then thermoformed in a sterile fashion, the presterilized product was added, and the lidding material was applied. A similar technique was used to produce a sterile surface on the lidding material, by stripping away part of a coextrusion. Thus no chemical or heat treatment was needed to sterilize the packaging material. The heat in the extruder during fabrication was sufficient to destroy any microorganisms present at that time, and the food contact surfaces were protected from subsequent contamination by the top layer of the coextrusion.

Current aseptic TFFS processes use hydrogen peroxide or steam to sterilize the food contact surface of the package.

4.13 Thermoforming Temperatures

To successfully thermoform materials, they must be heated to a temperature at which they can be deformed and retain their new shape. For amorphous polymers, the range of forming temperatures is generally fairly wide, making them easier to work with than crystalline polymers, which typically have very narrow forming ranges. For amorphous polymers, the forming temperature range generally starts about 20 to 30 °C above the polymer's glass transition temperature (T_g), although normal forming temperatures are 70 to 100 °C above T_g. Parts can be removed from the mold when they are cooled to approximately 10 to 20 °C below T_g, and mold temperatures are usually kept at 10 to 30 °C below T_g [1].

For crystalline polymers, the forming window is typically within a few degrees of the melt temperature. Solid phase pressure forming, mentioned earlier, involves forming at lower temperatures. Typical thermoforming temperatures for some of the major packaging plastics are shown in Table 4.1.

Table 4.1 Typical Thermoforming Temperatures for Packaging Plastics [1]

Polymer	Typical Forming Temperature (°C)
High density polyethylene	146
Polycarbonate	191
Polyethylene terephthalate	149
Polypropylene	154-163
Polystyrene	149
Polyvinyl chloride, rigid	138

4.14 Spin-Welding

Spin-welding is not a thermoforming method, but a type of heat sealing which can be used to assemble thermoformed halves into a bottle-shaped container. Thermoforming cannot ordinarily be used to produce bottles because of their shape. However, if you envision a simple cylindrical bottle cut in half, thermoforming can be used to produce the two halves. In spin-welding, the halves are combined into a container by rotating one half rapidly on top of the other (Fig. 4.13). The two pieces are designed with a slight interference fit, so heat is generated by friction when one half is rotated. A sensing device determines when the resistance to rotation indicates that the halves have fused together, and the container is then released and ejected. These containers cannot be formed with threaded finishes, because of the limitations of the thermoforming operation, so they are generally closed with membrane-type seals, usually of coated foil. A common material used in this process is impact grades of PS. Such containers have low cost compared to bottles. In addition, the two components can be nested, shipped compactly and economically, and then fabricated where they are filled. In variations of this process, thermoforming can be used to make one half of the container and injection or blow molding used to produce the other half.

4.15 Advantages and Disadvantages of Thermoforming

Thermoforming, as mentioned, is an economical way to make many types of packages. However, the shapes that can be produced are limited in comparison to the molding methods, which are discussed in Chapter 5. Except for matched-mold thermoforming, molding pressures are fairly low, and molds are relatively economical. Production rates can be quite high, but

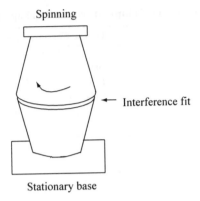

Spinning

Interference fit

Stationary base

Figure 4.13 Spin-welding of thermoformed containers

scrap rates are generally also high. The plastic must be formed into a sheet or film before thermoforming, except in melt-to-mold thermoforming. The thermoforming equipment can be located in-line with the sheet-making equipment and can utilize the residual heat in the sheet, but usually the sheet is fully cooled and rolled up, then unrolled and reheated for forming.

The basic thermoforming processes are limited to materials that deform easily when heated, yet have good melt strength. Other problems encountered in thermoforming include rupture of the sheet when too deep a draw is attempted or the forming temperature is too low; webbing at corners; wrinkling; warping; sheet sticking to the mold; and lack of forming detail. Proper control of sheet and mold temperatures and design of vacuum holes are very important in avoiding these problems. Thermoformed articles vary in wall thickness from one area of the article to the other, as discussed earlier, which must be taken into account when the package is designed to provide a certain degree of barrier protection for a product.

For many types of packaging, thermoforming is ideal. It is widely used in blister packages for a wide range of products. For containers, it is especially useful when relatively short production runs will not justify the expense of injection or blow molding.

References

1. Throne, J.L. (1987) *Thermoforming*, Munich:Hanser Publishers
2. Brody, A.L. and Marsh, K.S. (1997) In *Wiley Encyclopedia of Packaging Technology, 2nd ed.*, A. L. Brody and K.S. Marsh (Eds.), New York:John Wiley & Sons, pp. 914-921
3. Berins, M.L. (Ed.) (1991) *Plastics Engineering Handbook of the Society of the Plastics Industry*, 5th ed., New York:Chapman & Hall
4. Brody, A.L. and Marsh, K.S. (1997) In *Wiley Encyclopedia of Packaging Technology, 2nd ed.*, A. L. Brody and K.S. Marsh (Eds.), New York:John Wiley & Sons, pp. 910-914

5 Molded Packages

Complex packaging shapes such as bottles and caps that cannot be made by thermoforming are generally produced by injection or blow molding. In some cases, these methods are used even though thermoforming is a possibility, to achieve desired dimensional control or for other reasons.

5.1 Injection Molding [1-3]

Injection molding is commonly used for producing closures (caps and lids) of various types. It is also used for producing base cups for beverage bottles, margarine tubs, 19 L (5-gal) pails, and other containers and package components. Injection molding produces the preforms for injection blow molded bottles (see Section 5.3.2).

In injection molding, the plastic is melted in an injection molding machine, commonly an extruder much like those used for producing cast or blown film, except that it is equipped with a reciprocating screw that moves backwards and forwards in the barrel (Fig. 5.1). As the screw turns, melt is accumulated ahead of the screw, and the pressure forces the screw backwards in the barrel. When a sufficient volume of melt is accumulated, the screw stops rotating and is driven forward mechanically or hydraulically, injecting the plastic through a nozzle into the mold through the sprue, a system of runners, and finally into the cavity through the gate (Fig. 5.2). The mold is cooled, usually by water. When the package or package component is cool enough to maintain its shape without distortion, the mold opens along the mold parting line, and the object is ejected. The cycle time depends on the size of the molded part, as well as molding conditions, and can be as short as two seconds. Usually, the mold contains several mold cavities (sometimes as many as 64), and thus, several items are molded at the same time.

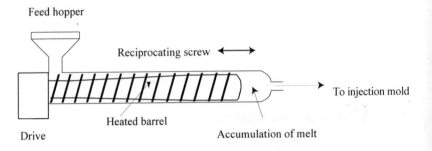

Figure 5.1 Injection molding machine

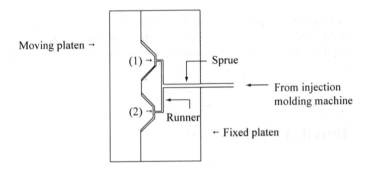

Figure 5.2 Injection mold, with mold cavities (1) and (2)

In such cases, it is very important to balance the flow of plastic so that the mold cavities fill equally. Generally, this means that the distance and geometry of each cavity from the sprue should be equal. Attention also needs to be paid to runner geometry to avoid dead spots. It is increasingly common for the runners to be heated (hot runner molding) so that the plastic in the runners does not solidify, and becomes part of the material filling the mold in the next cycle. Hot runner molding significantly reduces scrap and also improves productivity.

Packages produced by injection molding or injection blow molding can be recognized by the characteristic mark left at the gate where the plastic flows into the mold. It appears as a little nub, usually in the center of the object, and often surrounded by a slight depression called a sink mark.

When two-stage extruders are used, the plasticizing and injecting functions are separated. The extruder screw feeds the plastic into a separate chamber, known as the shooting pot, and then an injection piston forces the melt into the mold. Advantages of this system are that the extruder screw can rotate continuously, and greater accuracy in shot volume can be achieved. A third design, now nearly obsolete, involves a plunger, instead of a screw, which forces the plastic over a heated torpedo and into the mold. The poor homogeneity and mixing that result have led to the near total abandonment of this design in favor of the screw-type machines.

Injection molding machines are generally rated in terms of injection capacity and clamp force, two factors which are not directly related. Other important variables are L:D (length to diameter) ratio, barrel size, plasticizing rate, injection rate, and injection pressure. The major parts of the injection molding machine are the injection unit, the clamp unit, and the machine base which contains the power and control units. Clamp units can be either hydraulic or mechanical toggle types.

The injection molds themselves are generally made from high quality tool-grade steel, and are quite expensive. They may contain beryllium copper inserts to improve heat transfer rates for more rapid cooling. Wear plates are also incorporated, so that the molds can be more easily refurbished and have a longer life. The mold core is attached to the moving platen on the machine, and the cavity attached to the fixed platen. It is essential that the mold cores are accurately aligned with the cavities.

Venting must be provided to remove trapped air and prevent it from marring the finish of the package. Vent holes for this purpose are generally built into the mold parting line, where the core and the cavity come together. Large parts may require additional vents. The location of the mold gate is important in controlling the evenness of flow of the polymer into the mold, the polymer orientation in the finished part, and the presence or absence of weld lines and stresses. The gate is also designed to be thin in cross section, so that the plastic solidifies there quickly, sealing off the mold and allowing the screw to withdrawn without resulting in backflow.

Parts eject from the mold either mechanically or by air. For mechanical removal, stripper rings surrounding the core physically push the parts off the core. This is the method of choice for PS parts, which have a tendency to crack during the stress of air ejection. In air ejection, blasts of air loosen and blow the parts off the cores. It is used more frequently than mechanical ejection, since the system has fewer moving parts, thus requiring less maintenance and permitting more compact molds. Air ejection works better for thinner parts and parts can eject earlier in the cycle, because less sidewall strength in the component is required to avoid damage.

Obviously, the presence of significant undercuts, such as are present in a closure, complicates removal of the part from the mold. Undercuts are portions of a molded article which protrude into the side wall of the mold, thus preventing the object from simply sliding out of the mold (see Fig. 5.3) In some cases, closures or other packages with undercuts are flexible enough and undercuts small enough that the parts can be blown or popped loose from the mold. For threaded closures, however, this is frequently not the case, and more complex molds are required which either allow the closure to be unscrewed from the mold core, or are built with a collapsing core that retracts to free the closure.

Polymers used for injection molding typically have relatively low viscosities, and thus high melt flow rates, at the injection temperature used, to achieve the rapid mold-filling required.

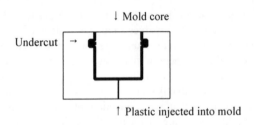

Figure 5.3   An undercut in an injection mold

Injection rates of up to 2 kg/s (4.4 lb/s) and injection pressures of up to 200 MPa (29,000 psi) are common [1]. The melt temperatures selected reflect balancing of the needs for low viscosity, rapid cooling, and avoidance of polymer degradation.

The temperature also has an effect on the orientation of the polymer in the finished part. Higher injection and mold temperatures, along with slower cooling, low injection temperatures, and thicker parts, decrease orientation. Orientation, as discussed previously, results in increased strength in the flow direction and decreased strength in the perpendicular direction. Of particular significance for injection molding, orientation also results in greater shrinkage on cooling in the flow direction than in the perpendicular direction. If molds are not designed correctly, this can lead to warping in the finished parts.

Injection molded packages and components are limited in the shapes which can be obtained, but they are characterized by excellent dimensional control. This is one of their primary advantages over thermoformed packages for applications like tubs where either molding method can be used. The ability to control the location of the plastic within the package can permit significant light-weighting of packages, and thus lead to better overall economics. For applications like threaded closures, thermoforming is not an option, and injection molding is the only practical method.

5.2   Closures [4,5]

Closures, or caps and lids, are a critical part of packaging systems. For containers, the closure is required to compensate for the manufacturing inaccuracies in the dimensions of both container and closure and still provide a tight hermetic seal, capable of protecting the container contents. Because of this requirement to maintain a tight seal, closures almost universally provide some type of resilient sealing surface in their design. While closures can be made of many materials, including cork, metal, and even glass, plastics are increasingly the closure material of choice. PP is the plastic most often used for closures as well as the most rapidly growing, followed by PS, HDPE, LDPE, and PVC in that order, with a small amount of thermoset phenolics and urea used as well [4]. Plastic closures can be classified into three major categories: friction, snap-on, and threaded.

5.2.1 Friction Closures

Friction closures seal by the friction between the outside of the closure and the inside of the container mouth (see Fig. 5.4). They are the plastic equivalent of the cork, and a common application is to seal inexpensive bottles of champagne. They are in very limited use, largely because they cannot compete in image with the more upscale natural cork, and often cannot compete functionally with other plastic closure designs.

Plastic "cork"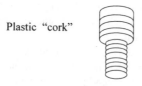

Figure 5.4   Example of a friction closure

5.2.2 Snap-Fit Closures

Snap-fit, or snap-on, closures are designed to deform as they pass over a protruding feature on the container, and then seal against the container by the resilience of the material exerting a force on the container as it attempts to return to its original dimensions (see Fig. 5.5). Snap caps and lids of various designs are widely used in packaging. This category includes the plastic lids on tubs of margarine, as well as the "line-up-the-arrows" type of child-resistant cap on pain reliever and prescription bottles and vials.

Snap-on closures are excellent for fast application. However, they are not suitable for containers with internal pressures exceeding one atmosphere, since the pressure might result in the closure snapping back off the container.

← Ring fits under retaining feature on container.

Figure 5.5 Snap-fit closure

5.2.3 Threaded Closures

Threaded closures are designed to screw on and off the container, or a screw base attached to the container, which is manufactured with a matching set of threads. These closures are generally termed continuous thread, or CT closures.. They can be used for packages which contain internal pressure, such as carbonated beverages, as well as for vacuum packages and those at atmospheric pressure. The sealing force provided depends on how far the closure is rotated on the package. Charts are available which give recommended torques for various sizes of containers [5]; several are presented in Table 5.1.

The removal force is generally less than the application force, unless there has been some interaction between the liner and the contents which acts to seal the liner to the container. Removal torque typically declines with time for the first several days to a month or so after application, because of stress relaxation and creep in the liner, closure, and/or container itself. Some recommended minimum removal torques are shown in Table 5.1, as well. As can be noted, recommended minimum removal torques are typically about half the recommended minimum application torque. Too low an application torque can result in containers which are not adequately sealed. If the torque is too high, consumers have difficulty removing the closure.

Closures are sized according to the nominal outside dimension of the container opening in millimeters plus a number which represents the style of finish. Both container and closure finishes are standardized so that, at least in theory, a closure of a given size and style should fit a bottle of that size and style, from any manufacturer. In the U.S., standards for closures are established by the Closure Manufacturers Association, standards for glass bottle finishes by the Glass Packaging Institute, and standards for plastic bottles by the American Society for Testing and Materials. Representative dimensions of closures are shown in Fig. 5.6 (see also Section 5.6). The most critical dimensions are T, the dimension of the root of the thread inside the closure; E, the inside dimension of the thread in the closure; H, the measurement from the inside top of the closure to the bottom of the closure skirt; and S, the vertical dimension from the inside top of the closure to the starting point of the thread.

To provide the resilience necessary to obtain a good seal, threaded closures often incorporate a liner system. The liner may contain plastic as the product contact layer, plastic foam or paperboard for resilience, and may also contain aluminum foil for barrier. Some closures are designed to incorporate a resilient sealing surface without requiring a separate

Table 5.1 Recommended Application and Removal Torques for Rigid Closures on Plastic Containers [5]

Cap Size mm	Application Torque N-m (in-lb)	Removal Torque N-m (in-lb)
15	0.7 (6)	0.3 (3)
24	1.1 (10)	0.6 (5)
33	1.7 (15)	0.8 (7)
48	2.3 (20)	1.1 (10)
70	3.2 (28)	1.6 (14)
120	5.4 (48)	2.7 (24)

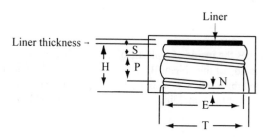

Figure 5.6   Standard closure dimensions

liner, and are known as linerless closures. Liners can be glued in place, but most often are held in place by being snapped behind a retaining ring built into the closure.

5.2.4 Specialty Closures and Associated Devices

A wide variety of specialty closures are available, along with various adjunct devices collectively known as fitments. Many of these are designed to dispense the product, such as pumps, sprays, shaker tops, spouts, and roll-on dispensers. Almost all are made wholly or primarily of plastic, and typically are injection molded.

Another set of devices provide evidence of tampering, including shrink bands on package necks, shrink wraps around containers, and tear-off rings on caps. Their function is to alert consumers to the fact that a package has been opened and may have been tampered with. These features are required by law on over-the-counter drug products in the U.S., and are increasingly found on packages for food and other products as well, although they are not required in these applications. Further discussion of tamper evidence is found in Section 1.7.4.

A third set of devices are those which make packages child-resistant. These features are designed to render access by young children difficult, to protect them from poisoning. Child-resistant packaging, known as "special packaging" in the relevant regulations, is required on most prescription drugs, aspirin, and other over-the counter drugs and household chemicals which pose a risk to children if accidentally ingested. Child-resistant packaging, along with current modifications to make it easier for the elderly to use those packages, is discussed further in Section 1.7.3.

5.3 Blow Molding

In blow molding, air pressure shapes the inside of a plastic object, much as is done in blown film, except a mold shapes the outside of the object. Blow molding is the only practical way to make plastic bottles and jars, and this method is also used for large plastic containers such as drums. There are two major types of blow molding, extrusion and injection blow molding.

5.3.1 Extrusion Blow Molding [2,3,6-8]

Extrusion blow molding is by far the most commonly used process for producing plastic bottles. It begins with the extrusion of a hollow plastic tube, just as is done for blown film, except that in this case, the extrusion is in a downwards direction. Two mold halves then close on the tube, capturing it as it is cut off from the extruder. A blow pin is inserted and air is blown into the mold, expanding the tube, or parison. In some cases the blow pin, cooled by water, assists in forming the inside of the finish. In other cases, the blow pin is inserted into a part of the molded object which is trimmed off in forming the final container shape, and the inside of the finish is formed only by air. The container is cooled by heat transfer with the mold, and when cool enough to maintain its shape, it is removed and the flash (excess material) trimmed from the container neck and bottom (Fig. 5.7). Generally, the flash is ground up and metered back into the extruder along with the virgin resin, although this is not always possible for heat-sensitive resins like PVC.

The extrusion of the parison can be continuous or intermittent. Continuous extrusion is preferred for most packaging applications because of its higher productivity. An additional advantage is that melt is not held up in the extruder, so thermal degradation is reduced. Intermittent extrusion requires a reciprocating extruder like those used for injection molding. It is commonly used for the production of very large blown containers, where a large parison must be produced in a very short time.

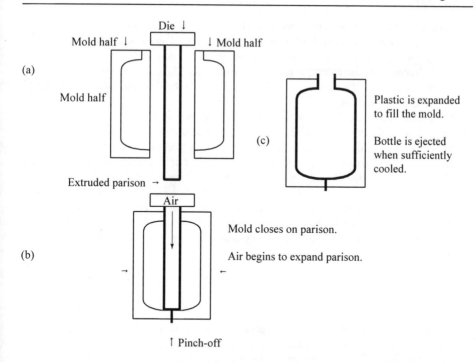

Figure 5.7 Extrusion blow molding

The parison diameter is usually nearly the same as the diameter of the container finish, although this depends strongly on bottle design, such as whether it has an offset neck or a handle, which require a wider parison. The ratio of the container diameter to the parison diameter generally should not be greater than four to one. The die dimensions are not identical to the parison dimensions, since stress relaxation and elastic memory cause the parison to shrink in length and swell in diameter and wall thickness.

Just as injection-molded containers can be identified by the mark left by the gate, extrusion blow molded containers can be identified by the marks left by the trimming of the flash. This is usually easiest to see on the bottom of the container, where it typically appears as a rough line running 2/3 or so of the distance from one side of the bottom to the other. Careful examination can also identify the roughness at the top of the bottle, on the top edge of the finish (the threaded part of the bottle).

Extrusion blow molding can be used to produce a wide variety of container shapes and sizes. It is particularly suited for production of bottles with handles and with offset necks, which cannot be effectively produced by injection blow molding. However, extrusion blow molding is not used very often for small bottles less than 200 ml (7 fl oz) in volume.

A single extruder can feed a multiple orifice blow molding die, producing multiple parisons simultaneously, which are then blown in multi-cavity molds. The blowing can be done at the same machine station as the extrusion, but it is most often done at a second station in the machine. Common machine designs include rotary and shuttle arrangements. In some cases, bottles are blown from the bottom instead of the top, so the bottle is produced upside down, known as rising mold designs.

Cylindrical bottles are the easiest to produce, but they are not always desirable because of the amount of shelf and shipping space they take up, and also for marketing reasons. An additional significant factor is that cylindrical containers tend to "panel-in" if any vacuum develops inside the container, as occurs if the container is hot-filled, or if there is any significant loss of product through the container walls. These distorted containers are likely to be rejected by consumers. A number of innovative designs have been developed to combat or disguise this distortion, but one of the simplest solutions is to use an oval container. The relatively flat side panel on an oval container can move inwards without producing any visible signs of distortion. However, the oval shape produces its own set of concerns, as discussed below.

5.3.1.1 Die Shaping

As was mentioned in the section on thermoforming, when the melted plastic is expanded by air pressure, it stretches and thins until it comes in contact with the mold, at which point it is cooled sufficiently that stretching stops. Thus, the parts of the parison which come in contact with the mold first are the thickest and the parts which have a greater distance to travel are the thinnest. For a cylindrical bottle, this is not a concern, since all parts of the bottle wall are approximately equal in thickness. However, if the bottle is significantly oval in cross section, a uniform parison produces a bottle with substantial differences in wall thickness. To make the bottle strong enough in its thinnest sections, some parts of the bottle wall are thicker than necessary, resulting in excess plastic used. To permit light-weighting of such containers, and resultant monetary savings, die shaping is routinely used.

Die shaping involves modifying the shape of the annular die so that the opening for plastic flow is not symmetrical. The parison produced does not have equal wall thickness on all sides. The thickest part of the parison is aligned with the part of the bottle farthest from the midline, thus producing more even wall thickness throughout the container (Fig. 5.8). The shaping can be done on either the die mandrel (the central part, core pin, or die pin) or the die bushing (the outer part).

5.3.1.2 Programmed Parison

A different problem occurs when the asymmetry in the bottle is vertical rather than horizontal, such as in bottles which have narrow waists or wide shoulders. A one-time change in the die

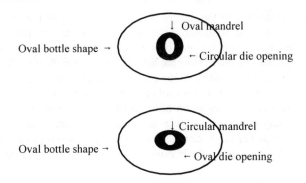

Oval bottle shape → ↓ Oval mandrel
← Circular die opening

Oval bottle shape → ↓ Circular mandrel
← Oval die opening

Figure 5.8 Die shaping can be accomplished by shaping either the mandrel or the die opening to meet the need for thicker or thinner sections of the parison to produce even wall thickness in the container

dimensions cannot produce even wall thickness in such containers. For these applications, a technique called programmed parison is used, in which the die opening is changed as the parison is produced, so that a thicker profile is produced in the part of the parison which forms the wider part of the bottle, and a thinner profile where the bottle is thinner (Fig. 5.9).

With a programmed parison, the die mandrel moves up and down inside the die bushing to increase or decrease the gap between the two. This process is also known as *dancing mandrel* blow molding. Sometimes the die bushing is moved instead of the mandrel. Die shaping and programmed parison techniques can be combined. The most sophisticated wall thickness control systems use deformable die rings so that both the mandrel motion and the die shape can be programmed to allow for extremely accurate control of the thickness profile of the extruded tube and hence of the blow molded container [6].

5.3.2 Injection Blow Molding [2,3,7,8]

Injection blow molding, a two-step process, is the other major bottle-forming process. In the first step, a preform, or parison, is produced by injection molding. In the second step, this preform is placed in a second mold and blown to produce the final container shape (Fig. 5.10). The injection molding of the preform is identical to injection molding processes described earlier, except that the mold has three parts, the two halves of the cavity and a core rod. Hot runner molding is almost universally used. The preform has a test tube-like shape, with the bottle finish complete at this point. Usually the preform is cooled only enough to allow molecular orientation to occur, and requires the support of the core rod to maintain its shape.

After the preform is conditioned to the desired temperature configuration, it is then transferred on the core rod into the container mold. Alternatively, the preform can be completely cooled, stored or shipped, and reheated before the final blowing step. Air at high pressure is introduced through the core rod to blow the bottle. A pressure higher than that used for ordinary blow molding is required because of the cooler temperature of the parison, and its resulting higher viscosity. When the bottle is cool enough to maintain its shape without distortion, it is ejected from the mold. Injection blow molding equipment is most often arranged in a three-stage rotary configuration, with an injection molding station, a blow molding station, and a cooling and ejection station. Bottles produced by this process can be identified by the mark left by the gate on the bottom of the bottle at the initial injection molding stage.

Extruded parison

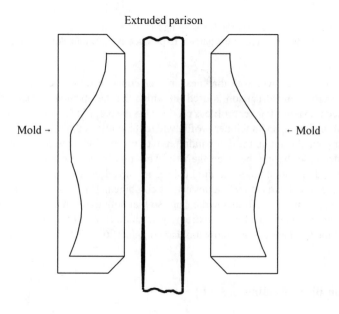

Mold →   ← Mold

Figure 5.9   Programmed parison

Typically, the diameter of the finished bottle should not be greater than three times the preform diameter, and the length of the preform should not be greater than ten times its diameter. The ovality of the container, defined as the ratio of the major diameter to the minor diameter, should not be more than two.

For small bottles, injection blow molding is generally preferred over extrusion blow molding, while extrusion blow molding is significantly more economical for larger bottles.

Mold costs for injection blow molding are higher than for extrusion, since two sets of molds are required, one for the preform and one for the final container. Another significant disadvantage is the extremely limited ability to produce handles on injection blow molded containers. Processes to produce small solid handles exist, though they are seldom used, but hollow handles cannot yet be made by injection blow molding [7].

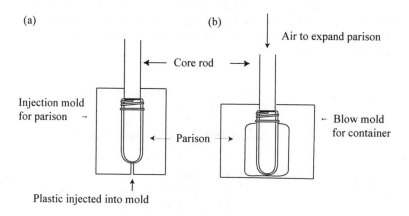

(a)   (b)

Air to expand parison

Core rod

Injection mold
for parison

Plastic injected into mold

Parison

Blow mold
for container

Figure 5.10   Injection blow molding, (a) injection molding of parison, (b) blow molding of container

The injection molded preforms or parisons usually have a wall thickness between 1 and 5 mm (40-200 mil). The ratio of the inside diameter of the parison to the inside diameter of the blown container is the *hoop ratio*, a factor that can be used to calculate the ratio of the wall thickness of the finished container to the wall thickness of the parison. The length of the parison is approximately 1 mm (40 mil) less than the inside length of the bottle in the blow mold, so there is no significant stretching in the vertical direction when the bottle is blown.

In general, injection blow molding produces higher quality containers than extrusion blow molding, especially in terms of finish tolerances. Extrusion can produce a wider variety of sizes and shapes, has higher production rates, and is more economical except for small bottles. Extrusion blow molding is preferred for PVC, while injection blow molding is preferred for PS and PET. HDPE and PP can be readily molded by either process.

In either process, the polymers in the containers have been oriented because of the radial stretching occuring in the blowing step. If it is desirable to maximize the mechanical or barrier properties of the material, the bottle can be biaxially oriented by stretching the bottle vertically as well as horizontally. This process is stretch blow molding, discussed next.

5.3.3   Stretch Blow Molding [3,7,8]

Stretch blow molding starts with the production of a preform, either by injection molding or extrusion blow molding (or, rarely, extrusion alone), which is shorter than the length of the finished bottle. The finish of the bottle is formed in this step. The preform, on a stretch rod, is conditioned to an accurate, consistent temperature, usually just above its T_g. The bottle finish is typically kept cool to avoid distortion. Then, the preform is stretched in the bottle mold by the vertical movement of the stretch rod while air is blown through the rod to expand the bottle into its final shape (Fig. 5.11).

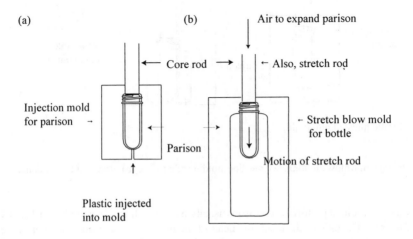

(a)   (b)   Air to expand parison

← Core rod →   ← Also, stretch rod

Injection mold
for parison   →

← Stretch blow mold
   for bottle

Parison

Motion of stretch rod

Plastic injected
into mold

Figure 5.11   Injection stretch blow molding, (a) injection molding of parison, (b) stretch blow molding of container

The resulting biaxially oriented container has improved impact strength, gas barrier, stiffness, clarity, and surface gloss. The process permits containers to have thinner walls, saving money through light-weighting. If the process begins with an extrusion molded preform, it is known as extrusion stretch blow molding; if the preform is produced by injection molding, it is injection stretch blow molding. Injection stretch blow molding is used for the vast majority of PET bottles, including all PET soft drink bottles. PEN bottles and some PP bottles are also produced by injection stretch blow molding. Extrusion stretch blow molding is used for some PVC bottles.

Stretch blow molding can be categorized as either a one-step or two-step process. In one-step stretch blow molding (also known as one-stage or in-line), the preform production, stretching, and blowing are all done in the same machine. The preform, after molding, is rapidly cooled to the stretch temperature and then immediately stretched and blown. In the two-step (two-stage or reheat-blow) process, the preform is completely cooled and then reheated to the stretch temperature before stretching and blowing. The primary advantage of the one-step process is energy savings. The primary advantage of the reheat-blow process is that the preform and the container can be molded at different times, in different places. The two-step process is widely used for beverage bottles, for example, so that the parisons can be produced in a central location, and the bottles blown at the filling point. This avoids the need to ship a lot of air, since the blown bottles are quite bulky. This system also means that the bottler does not need the expertise and equipment necessary for the injection blow molding part of the process.

For orientation to occur during stretching, the preform must be brought to a temperature within the orientation temperature range, below its crystalline melting point but above its T_g. For PET this temperature range is 88 to 116 °C, for PVC 99 to 116 °C, and for PP 104 to 127 °C [7]. The amount of orientation is affected by the temperature and the amount of stretching. The *axial ratio* is the ratio of the length of the stretched part of the bottle to the length of the stretched part of the parison. The finish is subtracted from both lengths in calculating the axial ratio since it is not stretched. The *blow-up ratio* is the product of the hoop ratio and the axial ratio. For maximum creep resistance, burst resistance and barrier properties, the blow-up ratio should be at least 10:1 and the hoop ratio at least 4.8:1 [3].

5.3.4 Molds for Plastic Bottles [3,8]

Blow molds for the manufacture of plastic bottles are commonly made of aluminum or beryllium copper. Beryllium copper has better heat transfer and resistance to wear, but is more expensive. Aluminum is most often used because it is significantly less expensive and provides good heat transfer, although it does wear more quickly. Blow molds are, in general, less rugged than injection molds, because they are not subject to pressures nearly as high as those in injection molding. Steel inserts are often used in the parts of the mold most subject to wear, such as the pinch-off in an extrusion blow mold (where the parison is pinched together to produce the bottom of the container). A mold typically can produce about 12 million containers, provided some refurbishing occurs.

Just as in injection molding, removal of trapped air is a concern, and venting must be provided, usually along the mold parting lines. Additional venting may be required, especially in large containers. Molds which are used to produce plastic bottles from HDPE, PP, and similar plastics which do not have very high clarity are often sandblasted to help reduce air entrapment. For clear bottles such as PVC, PS, and PET, sandblasting results in an unacceptable appearance, and the molds must be highly polished.

The cooling channels in the mold must be properly designed for the best productivity, since the cooling time is most often the controlling factor in cycle time. In addition to cooling by heat transfer to the mold, it is possible to cool the bottles internally with carbon dioxide, air, or a combination of air and water vapor. These methods are reported to increase production rates by as much as 50%, but many systems do not have the extruder capacity to support these increased rates, so they are not often used [8].

5.3.5   In-Mold Labeling [3,9,10]

Although still used on well less than half of plastic bottles, it is becoming increasingly common for bottles to leave the mold with their labels already in place, through a process known as in-mold labeling. This process involves placing labels in the mold while it is empty. The labels are held in place by a slight suction applied through the mold, and heat-seal to the bottle when it is blown. The labels themselves may be a heat-seal coated paper, but are increasingly likely to be a plastic compatible with the bottle, thus facilitating recycling. Plastic labels also generally look better than paper labels.

Advantages of in-mold labeling include the elimination of a subsequent labeling operation, the part of a packaging line often associated with the most down-time. The need to flame treat the bottles prior to labeling is also eliminated. In addition, since the label is nearly flush with the bottle wall, its print tends to be less scuffed, resulting in a better looking bottle. Claims have been made of improved strength and bulge resistance in the container, but others dispute these claims. Another advantage is that when in-mold labeling is coupled with in-case filling, the bottles can leave the bottle manufacturer, get filled and capped at the product manufacturer, and never leave their corrugated box.

The primary disadvantage is that the process is more complex than ordinary blow molding, and requires additional investment in capital equipment. It is not well suited to small production runs or where label copy changes frequently, even if bottles do not. In-mold labeling has been estimated to increase overall cycle time by 1.5 to 3.3 s [3], although others claim cycle time is not increased [9].

While most in-mold labeling is done on extrusion blow molded HDPE bottles, it has also been used on PET and PP stretch blow molded bottles. In-mold labeling of injection molded tubs is relatively common in Europe, but is just developing in the U.S.

5.3.6   Aseptic Blow Molding [8]

For some packaging applications, it is desirable to have a sterile bottle suitable for aseptic packaging. In aseptic packaging, a sterile product is sealed into a sterile container under sterile conditions, eliminating the need to sterilize the product-package system after filling (see Section 1.6). The heat of extrusion effectively sterilizes the plastic, but ordinary bottle manufacturing processes immediately expose the container to airborne microorganisms, so this sterility is lost. Aseptic blow molding avoids this by producing the bottle under sterile conditions.

The first step in the process is to do the blowing with sterile air. There are then two major variants of aseptic blow molding. In the first, the blow and hold process, the bottle is blown and sealed in the mold, and later at the filling line, cut open, filled, and sealed again. In the blow, fill and seal process, the bottle is blown, filled (often while still in the mold), and sealed, all in the same machine. This may be done at multiple stations in the machine, or all at the same station.

5.3.7 Heat Setting and Other Bottle Modifications

When containers cool, they often retain some thermal stress. If the containers are later heated, they can relax and undergo dimensional change. This is particularly a problem for containers which are to be hot-filled (filled with hot product, typically at temperatures of 77 to 93 °C [11]). While some bottles, such as PP, can undergo hot-filling without problem, PET bottles tend to distort severely at temperatures greater than 65° C. The major cause is the stretch blow molding process they have undergone.

Heat-setting processes have been developed to limit this distortion. One process uses a hot blow mold and provides sufficient residence time in the mold for stresses to relax and crystallinity to increase. The second method uses two blowing steps. The bottle is stretch blow molded into an initial form larger than the finished container, reheated in an oven causing it to undergo some shrinkage, and then re-blow molded in a second mold to produce the final shape. These bottles can withstand hot-fill conditions of 80 to 85 °C. In these processes, the crystallinity of the bottles is increased from the 20 to 28% achievable by stretch blow molding to 28% or higher [8].

Bottles designed for hot-filling must also withstand the partial vacuum generated inside when the product cools. This can cause the bottle to panel-in severely. Therefore, these bottles are designed to permit some distortion without detracting from the appearance of the container. Simply making the bottle oval rather than cylindrical can accomplish this, but a number of interesting designs have been developed for cylindrical bottles so that portions of the wall (called vacuum panels) can move inward as the product cools.

Another modification is production of a crystallized neck finish on PET bottles. Under normal molding conditions, crystallization of PET bottles is significantly retarded by rapid cooling, and the crystals formed are small and rodlike in shape. This allows the bottles to have good impact strength and to maintain their desirable transparency. However, this relatively amorphous PET is much less rigid than PET in its more crystalline form. When bottles are hot-filled, the finish area of the bottle may undergo some distortion. If the body of the bottle is kept cool while the finish is exposed to heat, the finish can be crystallized to a larger extent, with spherulitic crystals forming and rendering the material an opaque white. This more crystalline finish has increased capacity to maintain its dimensions during subsequent hot filling, and thus may provide a better seal.

5.4   Multi-Resin Bottles

Just as is the case for packaging films, there are applications where a bottle made of a single polymer cannot deliver the performance desired, or at least not at an acceptable price. In these cases, the use of a polymer blend or a multi-layer plastic bottle is often the solution.

5.4.1   Polymer Blends

Blends of plastics can provide properties different from those of a single resin. The plastics involved are fed into a single extruder, which does the blending, and the container is produced as usual. One example, mentioned previously, is a blend of nylon and HDPE, in which the nylon forms platelets in an HDPE matrix, forming a torturous path for hydrocarbon transmission, and thus improving the barrier properties of the container. A typical concentration is 5 to 18% nylon by weight. In another example, motor oil bottles are commonly produced from a blend of virgin and postconsumer recycled HDPE.

5.4.2   Coextruded Bottles [1,3,7,8]

Coextruded bottles are made by melting each resin in a separate extruder, feeding them into a die which forms them into a multi-layer tubular parison, and then blowing the containers as usual. As with coextruded film, the combining must be done carefully to avoid blending, with attention paid to melt viscosities and other factors. Also as in coextruded film, the use of a tie (adhesive) layer is often required to obtain adequate adhesion between resins of different types.

Two examples of coextruded bottle structures are shown in Fig. 5.12. The PP/EVOH catsup bottle was the first commercially successful bottle for an oxygen-sensitive food product in the U.S. The HDPE/regrind bottle structure is widely used for liquid laundry detergent and similar products. It permits incorporation of post-consumer recycled HDPE into a buried inner layer in a blend with regrind as the major structural component, while providing layers of virgin HDPE to protect the product against contamination and preserve the desired surface characteristics of the bottle.

Coextrusion is also used to produce opaque bottles with clear viewing stripes, so that the consumer can see the level of product inside the container. The unpigmented resin is processed in a small satellite extruder and introduced into the die as a separate vertical stripe in the parison. Such containers are used for motor oil, liquid laundry detergent, and other products.

(a)

PP
tie layer
EVOH
tie layer
regrind
PP

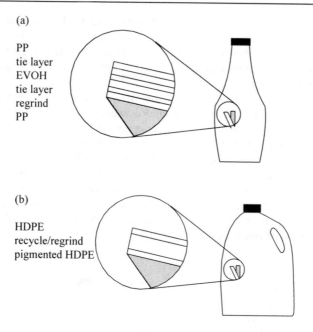

(b)

HDPE
recycle/regrind
pigmented HDPE

Figure 5.12   Typical coextruded bottle structures, (a) for oxygen sensitive foods, (b) for laundry products

5.4.3 Coinjection Blow Molded Bottles [7,12]

It was much more difficult to develop technology to produce multi-layer injection blow molded bottles than it was to produce multi-layer extrusion blow molded bottles. Containers made by coinjection blow molding are still rare, with Heinz, Inc.'s ketchup bottle the most common example in the U.S. Two techniques for producing the preforms for these bottles exist. One involves the simultaneous injection of two or more resins into the mold cavity. This is the technology used for the Heinz PET/EVOH ketchup bottle (Fig. 5.13). The second process involves a multiple set of preform molds, with a single resin injected in each step.

Once the multi-layer preform has been produced, the final blow molding stage is identical to ordinary injection blow molding. Stretch blow molding can also be applied, and, in fact, this is how the PET/EVOH bottles are produced.

As can be seen, the coinjection stretch blow molded PET bottles actually have a five layer structure, containing two layers of EVOH. Two thin EVOH layers are reported to provide better oxygen barrier than one thick EVOH layer, presumably because the effects of any defects in one layer are diminished by the presence of the second. It should also be noted that no tie layers are used in this structure, which means containers can be recycled at existing PET recycling facilities. Most of the EVOH is removed from the resin when the bottle is granulated and washed, and the remaining small residual does not present problems with the quality of the recovered material.

Recently Plastipak Packaging and Husky Injection Molding Systems introduced a third technology for making multi-layer PET bottles containing a barrier resin and up to 80% recycled PET. In this two-step process, a three-layer preform "liner" containing a barrier layer between two layers of virgin PET is first coextrusion blow molded. This liner is then placed by a robot into a conventional PET preform mold and over-molded with reclaimed PET. The preforms are then blown on standard reheat-stretch-blow systems. The claim is that a smaller quantity of the expensive barrier resin can be used because of better control over the layer thickness, and also that substantially higher levels of recyclate can be used. Other coinjection blow molding processes are limited to 35% recyclate. Bottles made by this process are expected to be available commercially by the end of 1997 [13].

PET
EVOH
PET
EVOH
PET

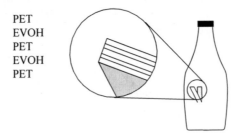

Figure 5.13   Coinjection stretch blow molded bottle

5.5   Surface Treatment [1,3,8,14,15]

The properties of plastic bottles can also be modified by surface treatment to enhance printability or label adhesion, or to improve barrier properties.

5.5.1 Flame Treatment

Flame treatment is used to improve the bottle's acceptance of printing ink or adhesive. It is difficult with polyolefin bottles, in particular, with their non-polar surfaces, to obtain adequate adhesion of inks or adhesives. By passing the bottle near a flame, some oxidation of the surface occurs, putting polar groups on the bottle surface and increasing its ability to bond. The flame treatment can also remove surface contaminants, such as mold release agents, which may interfere with bonding. The flame typically has a temperature of 1,090 to 2,760 °C, and contacts the plastic surface for less than 1 s [3].

5.5.2 Fluorination

Fluorination can improve the barrier capability of polyolefin containers, especially to hydrocarbons. The process involves exposing the polymer to fluorine gas, which reacts with the surface of the container, replacing some C-H bonds with C-F bonds. The polarity of the surface thus formed decreases the permeability of the container to nonpolar penetrants.

There are two basic approaches to the fluorination of bottles. The first is to blow the bottle with nitrogen containing a low percentage of fluorine. At extrusion temperatures, the fluorine reacts with the polyolefin, creating a fluoropolymer layer on the inside of the container. The second method involves normal production of the containers, and then placing them in a posttreatment chamber, where they are heated and exposed to fluorine gas in nitrogen. The polymer then reacts to form a fluorinated layer on both the inside and the outside of the container. In both cases, it is essential to purge and collect the highly corrosive fluorine gas and to ensure adequate protection for operators of the equipment.

Different levels of fluorination can be produced, depending on process conditions. In addition to improving barrier, fluorination also improves adhesion between the bottle and inks or adhesives (see Section 7.5).

5.5.3 Sulfonation

Another way to improve barrier properties of HDPE bottles to hydrocarbons is to use sulfonation. After molding, the bottles are treated with sulfur trioxide in an inert gas. This produces sulfonic acid groups on the bottle's surface, which are subsequently neutralized with ammonia or sodium hydroxide. As is the case with fluorination, sulfonation also improves the adhesion of inks and coatings. Proper procedures to ensure worker safety and avoid release of the sulfur trioxide and neutralization chemicals into the environment are necessary.

5.5.4 Coatings

Coatings of various types can also improve containers' barrier properties. PVDC coatings can increase barrier to hydrocarbons, water vapor, odors and flavors, and oxygen. It can be applied by a variety of methods, including spray and dip coating.

A more recent development is the use of silicon oxide-based coatings, primarily as oxygen barriers, generally on PET bottles. These function for bottles just as they do for films (see Section 3.4.1.2). The chemical vapor deposition process is used, forming an SiO_x film less than 2000 Å in thickness, which more than triples the oxygen barrier and improves water vapor barrier by 2 to 3 times [16].

5.6 Design of Plastic Bottles [5,17]

As is the case for closures, the design of plastic bottles is highly standardized, with the goal that any plastic closure of a designated size should fit any plastic bottle of matching size. ASTM issues design specifications for dimensions and tolerances of plastic bottles [17], as do other international standards organizations. Bottle finishes are sized by the nominal outside diameter in mm. The series indicates the finish style. Volume is specified by nominal overflow capacity. Thread profiles are of two main styles, the "L" or all-purpose thread, designed for plastic or metal closures, and the "M" or modified buttress thread, designed for plastic closures. The nomenclature for standard dimensions is illustrated in Fig. 5.14, and several standard dimensions are given in Table 5.2.

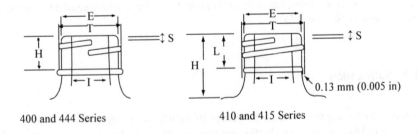

400 and 444 Series   410 and 415 Series

Figure 5.14   Standard finish dimension nomenclature for bottles

T refers to the diameter of the finish, including the threads. E is the outside diameter excluding the threads. I is the inside diameter of the finish. These dimensions are calculated by

Table 5.2 Standard Finish Dimension Tolerances for Plastic Bottles [17]

Series	Millimeter size	T, mm (in) min max	E, mm (in) min max	H, mm (in) min max	S, mm (in) min max	I, mm (in) min
SP-400	18	17.47-17.88 (0.688-0.704)	15.34-15.75 (0.604-0.620)	9.04- 9.80 (0.356-0.386)	0.56-1.32 (0.022-0.052)	8.25 (0.325)
SP-400	28	27.13-27.63 (1.068-1.088)	24.74-25.25 (0.974-0.994)	9.78-10.54 (0.385-0.415)	0.79-1.55 (0.031-0.061)	15.59 (0.614)
SP-410	18	17.47-17.88 (0.688-0.704)	15.34-15.75 (0.604-0.620)	12.90-13.66 (0.508-0.538)	0.56-1.32 (0.022-0.052)	8.25 (0.325)
SP-410	28	27.13-27.63 (1.068-1.088)	24.74-25.25 (0.974-0.994)	17.60-18.36 (0.693-0.723)	0.79-1.55 (0.031-0.061)	15.59 (0.614)
SP-415	18	17.47-17.88 (0.688-0.704)	15.34-15.75 (0.604-0.620)	15.29-16.05 (0.602-0.632)	0.56-1.32 (0.022-0.052)	8.25 (0.325)
SP-415	28	27.13-27.63 (1.068-1.088)	24.74-25.25 (0.974-0.994)	27.10-27.86 (1.067-1.097)	0.79-1.55 (0.031-0.061)	15.59 (0.614)
SP-444	28	27.13-27.63 (1.068-1.088)	24.74-25.25 (0.974-0.994)	14.53-15.29 (0.572-0.602)	4.06-4.83 (0.160-0.190)	15.59 (0.614)

taking the average of measurements across the major and minor axis of the container. The ratio of the two measurements is the ovality. Excessive ovality interferes with proper sealing. S is the distance from the top of the finish to the top edge of the leading thread. H is the distance from the top of the finish to the feature on the bottle that exceeds the T dimension and therefore stops the downward motion of the cap, the transfer bead in the 400 and 444 series, and the shoulder in the 410 and 415 series. In the second case, the L dimension is the distance from the top of the finish to the top of the transfer bead.

5.7 Plastic Tubes [18,19]

Plastic tubes are produced by a process that combines extrusion and injection or compression molding. Usually the body is produced by extruding a plastic tube. As the plastic emerges from the extruder, it is corona treated for ink adhesion, and then drawn over a chilled, internal forming mandrel and cold water is applied to the outside of the tube. As the tube cools, it shrinks to the diameter of the forming mandrel. The tube body, called a sleeve, is then cut to length. Coextruded tubes can be made, as well as single-resin tubes. Either before or after the sleeve has been printed and coated, the tube head is produced and attached to the body. The head itself can be produced either by injection molding or by compression molding. In one common method, the top of the sleeve is captured in an injection mold, and then plastic is injected, at relatively low pressures but high temperatures, to fill the cavity, forming the head and bonding it to the sleeve. In a finishing operation, the sprue is cut off the head and a cap applied. Another common method punches a disk out of a continuous hot strip of LDPE which adheres to the sleeve, and then uses compression molding to form the final head shape. A third method injects a ring of melted plastic in a female mold cavity and then compression molds the head. A process common in Europe uses a premolded head and spin-welds it to the tube.

It is also possible to extrude the tube and then immediately injection mold the head onto the tube, as part of the same operation. Blow molding can be used to form the tube in one piece, as a bottle, and transform it into a tube by cutting off the bottom.

Laminated tubes are typically formed from multi-layer material containing paper and plastic, and often aluminum foil as a barrier layer. The tube bodies are preprinted and then sealed into a cylinder. In this process, the edges of the sleeve are overlapped and compressed, squeezing some of the plastic out around the raw edges of the foil and paper. The tube is then cut to length, and next the head is molded and assembled to the body. The laminated body may contain as many as 10 layers, including the printing ink. When excellent barrier has been provided in the tube sleeve by aluminum foil, it is generally desirable to have good barrier in the head as well. This is commonly achieved by using a premolded insert of polybutylene terephthalate or of urea, inserted in the injection mold before the head is injection molded and fused to the sleeve.

After forming, the tube is printed and coated, if this has not already been done. The end of the tube is left open for filling. Closures on tubes are generally attached before the filling operation. The tubes are filled through the bottom and then sealed. Sealing for all-plastic tubes can be done using radiant heat, heated jaws, or ultrasonic sealing. Laminated tubes containing aluminum foil are often sealed using high frequency sealing. A growing method for sealing both laminated and all-plastic tubes is hot air sealing. In this process, the seal area inside the tube is heated with hot air and the tube is then pressed and chilled. Tubes can be sealed at rates of more than 100/min. in this fashion. The pressing operation for hot air or other types of filling is generally also used to apply a code to the tube, indicating its batch number, expiration date, and other information.

Finish dimensions for tubes are specified in a similar fashion to finish dimensions on bottles and the same basic nomenclature is used.

5.8 Rotational Molding

Rotational molding involves placing a plastic powder inside a hollow mold, closing the mold, and spinning it on two perpendicular axes simultaneously to distribute the powder throughout the mold by centrifugal force. The mold is then heated, still spinning, to melt the plastic. Spinning continues as the mold is cooled to solidify the plastic object. When it is sufficiently cool to maintain its shape, the mold is opened and the item removed. If the item is a container, an opening must then be cut into it. Rotational molding is not often used in packaging, although it has some uses for forming large bulk containers.

5.9   Advantages and Disadvantages of Molding

Injection molding can produce packaging components with exactly the dimensions desired. Because the mold cavity controls the shape of the finished part, the plastic is exactly where it is supposed to be if the mold is filled accurately. However, the shapes that injection molding can produce are limited because of the need to get the solid pieces of the mold out of the molded object. Designs such as collapsing core molds can help, but a bottle with a small neck (or even a large one) cannot be produced by injection molding. For items such as margarine tubs, which can be formed either by thermoforming or by injection molding, the latter may permit enough downgauging to more than pay for the higher production cost, because of the ability to more accurately control distribution of the plastic. The plastics used in injection molding must flow fairly rapidly to quickly fill the mold without "setting" the plastic as it cools, cutting off the path to some areas of the mold. Proper design of gates and consideration of flow patterns and weld lines where streams come together are essential to avoid stress and warping in the finished part. The injection molds themselves are relatively costly, as is molding equipment, largely because of the very high pressures they must withstand. Injection molding is the only practical method of making continuous thread closures, and is also widely used for lids of all types, as well as for tub and cup shapes.

Injection blow molding combines injection molding of a precisely formed parison with blow molding of the finished container. Thus, it is able to give fairly accurate control over container dimensions, especially in the critical finish area of bottles. Injection blow molding is more expensive than injection molding alone, since it requires two sets of molds and two molding processes. However, it is capable of producing shapes which cannot be produced by injection molding It also produces very little scrap. Injection blow molding can be used with resins without sufficient melt strength to be handled by extrusion blow molding. Injection blow molding is used for most PET bottles, as well as for most bottles used for pharmaceutical products. It cannot economically produce bottles with handles.

Extrusion blow molding is the simplest and generally the most economical process for making plastic bottles. Control over wall thickness is not as good as in injection blow molding, but can be enhanced by techniques such as parison programming and die shaping. It is capable of producing a very wide variety of bottle shapes, including bottles with handles, offset necks, dual chambers, and more. Considerable scrap is produced, especially with complex designs. Usually scrap can be reused in the process, but not always. Extrusion blow molding is also not very economical for very small bottles, so these are usually produced by injection blow molding. The resins used must have sufficient melt strength for the parison to remain unsupported, until it is captured in the mold. More than half of all plastic bottles are produced by extrusion blow molding.

Stretch blow molding can produce a bottle with biaxial orientation, thus improving its strength and barrier properties. It does require a two-step process, with very accurate temperature control of the parison so the distortion during stretching happens as desired, rather than some parts remaining too thick or too thin.

Molding is usually the method of choice for making rigid plastic containers, since thermoforming can produce only a limited selection of shapes. Rigid containers have the advantage over flexible containers of providing easier handling and dispensing, and often superior product protection, especially against impacts such as puncture and abrasion. Rigid containers are also self-supporting and can stand on a shelf and display product contents, often critical for sales. Their disadvantage is that generally more material is needed to produce a rigid package than a flexible package.

References

1. Carter, R. (1997) In *Wiley Encyclopedia of Packaging Technology, 2nd ed.*, A.L. Brody and K.S. Marsh (Eds.), New York:John Wiley & Sons, pp. 503-511
2. *Modern Plastics Encyclopedia Handbook* (1994) New York:McGraw-Hill
3. Berins, M.L. (Ed.) (1991) *Plastics Engineering Handbook of the Society of the Plastics Industry*, 5th ed., New York:Chapman & Hall
4. Brody, A.L. and Marsh, K.S. (1997) In *The Wiley Encyclopedia of Packaging Technology, 2nd ed.*, A.L. Brody and K.S. Marsh (Eds.), New York:John Wiley & Sons, pp. 206-220
5. Hanlon, J.F. (1992) *Handbook of Package Engineering*, 2nd ed., Lancaster, Pa:Technomic Pub. Co, Inc.
6. Fritz, H.G. (1988) In *Plastics Extrusion Technology*, F. Hensen (Ed.), Munich:Hanser Pub., pp. 363-429
7. Rosato, D.V. (1988) In *Blow Molding Handbook*, D.V. Rosato & D.V. Rosato (Eds.), Munich: Hanser Pub., pp. 1-96
8. Irwin, C. (1997) In *The Wiley Encyclopedia of Packaging Technology, 2nd ed.*, A.L. Brody and K.S. Marsh (Eds.), New York:John Wiley & Sons, pp. 83-93
9. Sneller, J. (1986) In *The Wiley Encyclopedia of Packaging Technology*, M. Bakker (Ed.), New York: John Wiley & Sons, pp. 238-239
10. Schultz, R.B. (1997) In *Wiley Encyclopedia of Packaging Technology, 2nd ed.*, A.L. Brody and K.S. Marsh (Eds.), New York:John Wiley & Sons, pp. 297-298
11. Bourque, R.A. (1997) In *Wiley Encyclopedia of Packaging Technology, 2nd ed.*, A.L. Brody and K.S. Marsh (Eds.), New York:John Wiley & Sons, pp. 492-496
12. Nakamura, Y. (1988) In *Blow Molding Handbook*, D.V. Rosato & D.V. Rosato (Eds.), Munich: Hanser Pub., pp. 117-148
13. Mapleston, P. (Feb. 1997) Modern Plastics, p. 33

14. Brody, A.L. and Marsh, K.S. (1997) In *The Wiley Encyclopedia of Packaging Technology, 2nd ed.*, A.L. Brody and K.S. Marsh (Eds.), New York:John Wiley & Sons, pp. 864-867
15. Finson, E. and Kaplan, S.L. (1997) In *The Wiley Encyclopedia of Packaging Technology, 2nd ed.*, A.L. Brody and K.S. Marsh (Eds.), New York:John Wiley & Sons, pp. 867-874
16. Eastapac Company (1990) *Plasma Enhanced Barrier System*, Kingsport, TN:Eastapac Co.
17. ASTM (1988) *Standard Specification for Dimensions and Tolerances for Plastic Bottles*, D2911-82, Philadelphia, Pa:ASTM
18. Stewart, D.A. and Brandt, R. (1997) In *The Wiley Encyclopedia of Packaging Technology, 2nd ed.*, A.L. Brody and K.S. Marsh (Eds.), New York:John Wiley & Sons, pp. 941-945
19. Miller, T. and Boedekker, T.J. (1997) In *The Wiley Encyclopedia of Packaging Technology, 2nd ed.*, A.L. Brody and K.S. Marsh (Eds.), New York:John Wiley & Sons, pp. 939-941

6 Foams, Other Cushioning Materials, and Distribution Packaging

6.1 Introduction

Plastic foams are commonly used in packaging as cushioning materials, and are also used in the form of containers. The most commonly used packaging foam has long been PS, but packaging foams also include urethanes, PE, PP, and other materials. They have in common a very light weight, good insulating capacity, and the ability to absorb shocks and protect the enclosed product [1]. Their most important use is protection of the product during distribution

Foam cushioning can be utilized as molded pieces, or as loosefill. Molded pieces are most commonly used for large heavy objects, while loosefill is suited primarily for light weight objects. An alternative is foam-in-place systems, where the foam is formed within the shipping container, tightly fitting around the packaged object.

Foams are classified as open cell if there are communicating passages between adjacent cells, or as closed cell if the cells are surrounded by a polymer matrix. A sponge is a common example of an open cell foam. Most packaging foams are closed cell. Closed cell foams are much less absorbent than open cell foams, a significant advantage for most packaging applications.

Other cushioning materials can also provide product protection. The selection of an appropriate cushion requires knowledge about the cushioning material properties, the product characteristics, and the characteristics of the distribution system.

Pallets, drums, bins, totes, etc. are also important parts of the distribution system. While corrugated fiberboard boxes and wood pallets are the most common distribution packages, plastics are used increasingly because of their longer life, easy cleaning, and other important characteristics.

6.2 Polystyrene Foam [1-6]

PS foam is widely used in both molded shapes and extruded form. Molded shapes are commonly known as expanded PS while the extruded material is known as extruded PS foam. PS foam is relatively inert and is acceptable for use in food packaging applications. This material is often incorrectly called "styrofoam," which is actually a Dow Chemical Company trade name for building insulation and not a packaging material.

Cushioning materials are important applications for PS foam, but they have a variety of other uses as well. They are widely used for meat and produce trays and as insulation for products which must be kept either hot or cold. These uses range from shipping for sensitive pharmaceuticals to clamshells for fast food. PS foam labels on glass bottles for carbonated beverages are effective in reducing surface abrasion on the containers, permitting significant light-weighting of the containers and rendering them more competitive with aluminum cans and PET bottles. Non-packaging uses, such as cafeteria trays, disposable dinner plates, and coolers are also common, as well as the ubiquitous disposable coffee cup.

6.2.1 Expanded Polystyrene Foam (EPS)

EPS is prepared by impregnating PS granules or beads with a hydrocarbon blowing agent, usually pentane, in amounts less than 8% by weight. The granules are then heated to 85 to 96 °C (185 to 205 °F), vaporizing the pentane and creating internal pressure which expands the beads. Typical bead expansion is 25 to 40 times the original size, to a density of 16 to 26 mg/cm^3 (1 to 1.6 lb/ft^3) [1]. These pre-expanded beads are aged to reach equilibrium, and are then packed into a mold, where they are held under several tons of pressure while steam is introduced directly into the mold. The heat and pressure cause the beads to fuse together, producing a semirigid closed cell foam. The object is held in the mold for a cooling cycle, and then released when it is dimensionally stable.

The beads are manufactured in three sizes, small, medium, and large. The choice of the bead size is determined by the wall thickness of the molded part, with small beads necessary for thin walls. The amount of expansion of the beads is controlled by the amount of blowing agent and the time and temperature of expansion. EPS is typically the least expensive packaging foam available, and is therefore the material of choice if it can perform acceptably.

6.2.2 Extruded Polystyrene Foam

As its name indicates, extruded PS foam is prepared by extrusion, rather than by molding. The polymer is melted in an extruder, the blowing agent and a nucleator are mixed in, and the blend is extruded. The nucleator provides foaming sites to help obtain the desired cell size and uniformity. Talc, citric acid, and blends of citric acid with sodium bicarbonate are common choices.

The blowing agent is usually injected into the melt in the form of a liquid, although it is sometimes introduced as a pressurized gas. Hydrocarbons are the most common blowing agents, but use of carbon dioxide is growing. Chlorofluorocarbons were commonly used until their connection to ozone depletion was discovered, but their phase-out is now nearly complete. Hydrofluorocarbons, which contain no chlorine and therefore do not deplete ozone, are also used in some places, but this use is prohibited in the U.S. Blends of blowing agents are also available.

The melt is kept under pressure until it leaves the die. Release of the pressure vaporizes the blowing agent, and the melt expands. To obtain sufficient melt strength to produce a good foam, it is necessary to cool the melt after the blowing agent has been introduced. This can be accomplished within the original extruder (single or twin-screw) by incorporating a cooling zone in the design, but is most often done by using a two extruder system, known as a tandem system. The cooling takes place in the second extruder which is fed with the melt from the first extruder. In this design, the plastic and nucleating agent are fed into the first extruder, the blowing agent introduced about two-thirds of the distance down the extruder, and the mix passed through a screen-changer for filtering before it is fed into the second extruder.

An annular die is generally used to form the foam. A compressed air ring around the circumference of the die forms a thin skin on the foam surface as it foams on exiting the die. The tubular foam then passes over a water-cooled mandrel for further cooling. It is next slit, flattened, and wound into rolls. Slit dies can also be used, producing a flat sheet, but use of annular dies is much more common.

The density of the foam is controlled by the amount of blowing agent which is introduced. The size and number of cells are controlled by the amount of nucleating agent. PS foams used for packaging typically have densities in the 60 to 100 kg/m^3 (4 to 6 lb/ft^3) range.

Thermoforming, most often using matched molds, is commonly employed to convert the sheet into its final shape. For best performance, the sheet should be aged for 3 to 5 days before thermoforming to permit the cell gas pressure to reach equilibrium. The scrap from thermoforming can be ground and densified in an extruder for recycling. Used materials can also be recycled.

6.2.3 Styrene Copolymer Foams

Copolymers of styrene and other monomers can also be used as the basis for foams. One of these is styrene-acrylonitrile (SAN) foams. This produces semirigid foams with better performance than EPS in repeated drops and for heavy products (high static loads). The density of SAN foam is usually about 16 kg/m³ (1 lb/ft³). Prices are higher than for EPS.

6.3   Other Plastic Foams

Virtually any polymer can be made into a foam. While PS remains the most commonly used packaging foam, several other materials have significant use.

6.3.1 Polyolefin Foams [1-3,6]

PE foams are, like polystyrene, available in two varieties, expanded (generally termed moldable) and extruded. Costs are higher than for PS. The foams are more flexible, and better able to provide protection from multiple impacts. Cross-linked grades are also available. PE foams have densities in the 16 to 32 kg/m³ (1 to 2 lb/ft³) range for use as an overwrap in transportation of goods [1,2]. PP foams are also available, with properties similar to PE foams, but with somewhat greater rigidity.

   A 50% PE/50% PS copolymer foam, termed expanded PE copolymer (EPC) is manufactured by ARCO Corp. and tradenamed Arcel. Its characteristics are generally between those of PE and PP, but exceed either material in toughness.

   The manufacture of extruded polyolefin foams is very similar to manufacture of extruded PS foam, with an annular die and forming mandrel used for sheet products. Extrusion through rectangular dies onto a conveyor belt is also used to produce planks. Blowing agents are typically hydrocarbons or blends of hydrocarbons and carbon dioxide.

   Crosslinked PE foams are produced somewhat differently. Instead of using physical blowing agents like hydrocarbons and carbon dioxide, chemical blowing agents such as azodicarbonamide are used. The resin, additives, crosslinking agents, and blowing agents are mixed together at temperatures below the activation temperature of the blowing agent, and extruded into a flat sheet or other profile. Crosslinking is then accomplished, either chemically or by radiation. Radiation is typically used for thin materials, and chemical cross-linking for thick profiles. Finally, the foam is expanded by exposure of the cross-linked material to hot air (about 200 °C) to activate the blowing agent.

6.3.2   Polyurethane Foams and Foam-in-Place Systems [2-3]

Polyurethane foams are most often used in packaging as part of a foam-in place operation, in which the foams are produced inside of the shipping container. In this technique, a mixture of a polyol and an isocyanate, with other ingredients, is injected into the erected shipping container. As the foaming reaction begins, a PE film is placed over the top of the mixture, the product is added, and then a second layer of film is placed over the product. Another shot of the polyurethane mixture is added and the case is quickly sealed. The foaming continues, tightly encapsulating the product in a nest within the foam (Fig. 6.1). The polyethylene sheets protect the product from direct contact with the foam.

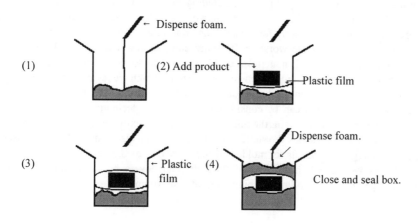

Figure 6.1   Foam-in-place packaging system

Generally for high volume applications, foam-in-place systems are not economically advantageous compared to use of molded cushioning. They are, however, often the method of choice in low volume applications, particularly where a large variety of different shapes and sizes of products are packaged, especially if these objects are relatively heavy so that loosefill cushioning and other alternatives do not work well.

Polyurethane foams can be formed using a fixture rather than the actual product, to produce cushions, encapsulated in polyethylene sheet, of the desired size and shape. These cushions are then used just like molded cushions made from other plastic foams. Foams can also be produced in bags rather than directly in the box. Polyurethane foams are produced by extrusion methods, as well.

An interesting new development designed to be competitive with polyurethane-based foam-in-place systems uses adhesive-coated expanded PS shapes. The shapes are deposited into the carton, covered with a release film, the product added, another sheet of release film put in place, and the carton filled with the adhesive-coated shapes. When the water-based adhesive dries, the result is a pair of resilient cushions surrounding the product, just as foam-in-place does. It is claimed that the adhesive-coated EPS shapes can be recycled with other EPS products.

6.3.3 Starch-Based Foams

Loosefill polystyrene "peanuts" and other shapes have been regarded with disfavor by environmentalists for a variety of reasons, including the association of polystyrene foams with chlorofluorocarbons (which, as has been discussed, are no longer used for this purpose), solid waste concerns, and, particularly for loosefill material, litter. The tendency of loosefill PS to be a litter problem is made worse by its static cling characteristics, as anyone who has unpacked something with this material as cushioning has no doubt observed. These problems led to the development of alternative cushioning systems which are regarded by some as more environmentally friendly. One of these is starch-based loosefill packaging shapes. While it can be argued whether starch can be considered a plastic, the technique for producing these materials does involve plasticizing the starch with water in an extruder and molding the shapes. Not only do these shapes lack the static cling of PS, they are also water-soluble, so disposal is easier. Analogous materials for molded cushions are also available.

One disadvantage of these materials is their tendency to be susceptible to moisture sorption. They may even collapse in high humidity environments.

6.4   Other Cushioning Systems

Two major alternatives to the use of plastic foams as cushioning materials also rely on the use of air to provide resiliency. One of these is bubble wrap, in which air bubbles of a defined size and pressure are sealed between two plastic sheets. These materials are exceptionally light weight, with densities as low as 11 kg/m^3 (0.7 lb/ft^3). They are used in the form of bags and envelopes, as well as wrapping material. Bubbles are available in several different sizes, with larger sizes intended for heavier duty applications.

Taking this concept one step further, transparent inflatable bags can also be used to provide a cushion of air around the product, while taking up minimal space during shipping and storage prior to use.

6.5   Design of Molded Cushioning Systems [7]

Selection of an appropriate cushioning system to protect a product during transportation requires knowledge of the product fragility, the distribution hazards, and the cushion characteristics. Product fragility is generally determined by testing, and reported in "G-levels." As discussed in Section 1.5, rules of thumb exist for distribution hazards, depending on the product weight and the handling system. These generally translate into a "drop height" that the product can be reasonably expected to encounter and therefore should be protected against.

The next step is information about the cushion characteristics. Manufacturers of foam cushioning make available characteristic "cushion curves" for their materials, which relate the average deceleration to the static loading as a function of drop height, cushion thickness, temperature, and number of impacts (see Fig. 6.2). There are also computer programs which contain mathematical models of cushion curves for various types of foams, sometimes along with cost information.

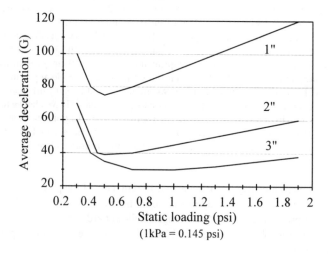

Figure 6.2   A typical cushion curve for a hypothetical packaging foam, for drops 2 to 5 from 91 cm (36 in), for 2.54 cm (1 in) thick, 5.08 cm (2-in) thick, and 7.62 cm (3-in) thick foam

Combining these two sets of information allows selection of an appropriate cushion thickness and load-bearing area which can reasonably be expected to provide adequate protection.

Transmission of vibrations to the product is also important, and characteristic vibration transmissibility plots for foams are also available. Of course, the design process should be followed by the preparation of a model package and suitable testing of the product/package system, or, in case of a very expensive product, of a mock-up, using accelerometers, to determine actual performance.

The process of determining the required cushioning for a product is illustrated in the following example:

Example: For the cushion with a characteristic cushion curve shown in Fig. 6.2, calculate the load-bearing area which will provide adequate protection for a product measuring 30.5 cm (12 in) square, weighing 11.4 kg (25 lbs), and with a fragility of 50 G.

Solution:
From Table 1.2, the maximum expected drop height for a product with this weight is 91 cm (36 in). Thus we can use the cushion curve in Fig. 6.2. We can eliminate the 2.54 cm (1 in) cushion since it does not provide decelerations below 50 G at any static loading. However, the 5.08 cm (2 in) and 7.62 cm (3 in) cushions are possibilities. Next we need to calculate the static loading. From the 5.08 (2 in) cushion curve, the static loading must be between 2.8 and 10 kPa (0.4 and 1.5 psi) to keep the G under 50. For the 7.62 cm (3 in) cushion, the static loading must be above 2.4 kPa (0.35 psi) and can be as high as 13.1 kPa (1.9 psi), the maximum extent of our information. A 11.4 kg (25 lb) object will give a loading of 2.8 kPa (0.4 psi) with a cushion area of 403 cm² (62.5 in²) and a loading of 10 kPa (1.5 psi) with a cushion area of 108 cm² (16.7 in²). Since the footprint of our item is 929 cm² (144 in²), either of these or anything in between is feasible. Thus our cushion can provide any contact area between 108 and 403 cm² (16.7 and 62.5 in²), if we are using a 5.08 cm (2 in) cushion, but should not provide more or less than this to get adequate performance. If we choose to use a 7.62 cm (3 in) cushion, by the same process we determine that the cushion area should be at least 85 cm² (13.2 in²) and not more than 461 cm² (71.4 in²) to be ensured of adequate performance. The precise cushion geometry can be anything that is practical for the particular product.

The fragility of the product may be different in different parts of the product. For example, if the product falls on its side as opposed to its bottom, there may be greater or less damage. In general, at least for products contained in corrugated fiberboard boxes, flat drops are more damaging than corner drops, because when the box deforms during a corner drop, it absorbs some of the impact energy.

6.6   Plastic Pallets [8]

Plastic pallets are increasingly replacing wooden pallets in product distribution. Although the cost of a plastic pallet can be four times the cost of a wooden pallet, its lifetime is usually much longer, often resulting in overall cost savings. While repair of plastic pallets is not often feasible, they can usually be readily recycled when they reach the end of their useful life, which is typically five to nine years.

Most plastic pallets are manufactured from HDPE. PS, PP, and fiberglass-reinforced plastics have also been used. Steel reinforcements can be added to plastic pallets to increase their load-bearing capacities.

Plastic pallets are often manufactured by structural foam molding. In this process, low pressure injection molding produces pallets with a solid skin and a foamed core. Wall thickness is generally 4.8 to 25.4 mm (0.19 to 1.0 in). Ordinary injection molding is also used. Walls are thinner, typically 0.8 to 9.5 mm (0.03 to 0.375 in). Rotational molding is used occasionally, producing pallets with 3.2 to 6.4 mm (0.125 to 0.25 in) walls. Thermoforming, especially twin sheet thermoforming, is another common forming method, with wall thicknesses also in the 3.2 to 6.4 mm (0.125 to 0.25 in) range. Polyurethane pallets can be produced by reaction injection molding (RIM), and have walls from 3.2 to 51 mm (0.125 to 2.0 in) in thickness.

Plastic pallets are most suitable for use where the distribution environment is relatively controlled, so that loss does not present a large problem. Obviously, to justify the initial higher cost, it is necessary to ensure that most of the pallets have a long service life. Some designs for plastic shipping containers combine the functions of pallet and container, and are often collapsible for economy in storage and shipping when empty.

6.7 Plastic Drums and Other Shipping Containers [9-12]

Plastic is increasingly being used for drums, pails, totes, and similar shipping containers. The resin of choice is usually high molecular weight HDPE. Plastic drums are most often used to ship various chemicals, but they are also frequently found in the food-processing industry. Standard drum sizes in North America are 57, 76, 114, 132, 208 and 216 l (15, 20, 30, 35, 55, and 57 gal). In Europe and most of the rest of the world, standard sizes are 30, 60, 120, 210 and 216 l (7.9, 15.9, 31.7, 55.5, and 57 gal).

Drums are available in closed-head (or tight-head) design, as well as in open-top designs. Open-top drums are used mostly for water-based products, and can generally be easily cleaned and reused. Plastic drums can be used for the shipment of hazardous materials provided the appropriate testing has been done and the containers are properly labeled (see Section 1.7.2).

An older alternative to plastic drums is steel drums containing a polyethylene liner. The liner adds chemical resistance and simplifies cleaning and reuse of the drum, since the liner can be discarded and replaced with a new one.

Plastic pails are available in sizes from 4 to 23 l (1 to 6 gal). Open top containers are referred to as pails, and account for about 75% of the market, while closed-head containers are called either pails or jerrycans in North America and jerrycans in most of the rest of the world. They generally are equipped with a handle. Most are made of HDPE containing a butene comonomer. As with drums, pails used for shipping hazardous materials must have passed the required performance tests. Pails are generally manufactured by injection molding.

Plastic crates and boxes are available in a variety of sizes and designs. They are typically injection molded, usually from HDPE in the U.S., but often from PP in the U.K. and some other parts of the world. Most are designed to be reusable, and many have the ability either to collapse or to nest to minimize space requirements on the return trip.

An alternative material for boxes and crates is plastic corrugated. PP copolymer is used most frequently, with HDPE also reasonably common. Polycarbonate is used for some special applications. The board can be made from extruded profiles, or by laminating three separate sheets together. Boxes can be made by die-cutting, scoring, and folding the board, much as is done for ordinary corrugated fiberboard. Joints are most effectively made by using high frequency welding. Silicone or hot melt adhesives have been successfully used if the board has been corona-treated. Metal staples (stitching) can also be used, but create weak spots in the container in the area of the staples. Plastic corrugated is particularly suitable for conductive containers for electronic parts that need protection against electrostatic discharge. The conductive material can be introduced into the polymer before extrusion, reducing the potential for wear and fiber generation, which were significant problems with carbon films printed onto corrugated paperboard, the older alternative. Plastic corrugated has also been widely used by the U.S. Post Office.

6.8 Packaging for Electrostatic Discharge Protection [13]

Sensitive electronic products require special attention to protect them from electrostatic discharge (ESD). For unmodified plastics, ESD is a significant problem, since the electrical nonconductivity of plastics makes them easily susceptible to the accumulation of static charges. Thus, when plastics are used to package sensitive devices, they must be modified in some fashion to make them less susceptible to the build-up of static charges, and to provide some way to dissipate the charges which are created. The way to do this is to provide a conductive path.

The most common way to alter the plastic is to add some component which attracts and retains a thin layer of surface moisture, at the same time weakly ionizing it. The moisture then provides the needed conductive path for dissipation of the static charge. The additive can be applied to a surface layer, can be distributed through the bulk of the material, or can even be confined in an inner layer of a coextruded material. Typical additives include ethoxylated amines, quaternary amines, and ethoxylated amides.

Antistatic materials are available in a variety of forms, including film, foam, bubble wrap, and containers. ESD films are also available in heat-shrinkable form to minimize static generation by friction within the bag, as well as transmission of static charges from outside the bag.

ESD protective materials are often tinted pink as a visible signal of their antistatic properties. In fact, the first military specifications for these types of products required visible identification, resulting in the adoption of the pink coloration.

In addition to additives, it is possible to employ plastic materials with inherently static-dissipative properties. These are still largely in the experimental stage. One promising material is polyamide ethylene oxide block copolymers. Costs of these materials tend to be higher than materials containing additives in relatively inexpensive resins like polyethylene. Use of conductive fillers in plastics has also been investigated, but has not been very successful.

References

1. Rodgers, P. (1986) In *The Wiley Encyclopedia of Packaging Technology*, M. Bakker (Ed.), New York: John Wiley & Sons, pp. 341-345
2. *Modern Plastics Encyclopedia Handbook* (1994) New York: McGraw-Hill
3. Berins, M.L. (Ed.) (1991) *Plastics Engineering Handbook of the Society of the Plastics Industry*, 5th ed., New York:Chapman & Hall
4. Wagner, P.A. (1997) In *The Wiley Encyclopedia of Packaging Technology, 2nd ed.*, A.L. Brody and K.S. Marsh (Eds.), New York:John Wiley & Sons, p. 449.
5. Brody, A.L. and Marsh, K.S. (1997) In *The Wiley Encyclopedia of Packaging Technology, 2nd ed.*, A.L. Brody and K.S. Marsh (Eds.), New York:John Wiley & Sons, pp. 370-378
6. Suh, K.W. and Tusim, M.H. (1997) In *The Wiley Encyclopedia of Packaging Technology, 2nd ed.*, A.L. Brody and K.S. Marsh (Eds.), New York:John Wiley & Sons, pp. 451-458
7. Brody, A.L. and Marsh, K.S. (1997) In *The Wiley Encyclopedia of Packaging Technology, 2nd ed.*, A.L. Brody and K.S. Marsh (Eds.), New York:John Wiley & Sons, pp. 287-293
8. Luft, L.T. (1986) In *The Wiley Encyclopedia of Packaging Technology*, M. Bakker (Ed.), New York:John Wiley & Sons, pp. 488-492
9. Poetz, B. and Wurzer, E. (1986) In *The Wiley Encyclopedia of Packaging Technology*, M. Bakker,(Ed.), New York:John Wiley & Sons, pp. 247-249
10. Brody, A.L. and Marsh, K.S. (1997) In *The Wiley Encyclopedia of Packaging Technology, 2nd ed.*, A.L. Brody and K.S. Marsh (Eds.), New York:John Wiley & Sons, pp. 704-708
11. Brasington, R.M. (1997) In *The Wiley Encyclopedia of Packaging Technology, 2nd ed.*, A.L. Brody and K.S. Marsh (Eds.), New York:John Wiley & Sons, pp. 110-112
12. Brody, A.L. and Marsh, K.S. (1997) In *The Wiley Encyclopedia of Packaging Technology, 2nd ed.*, A.L. Brody and K.S. Marsh (Eds.), New York:John Wiley & Sons, pp. 285-287
13. Havens, M. and Fowler, S. (1986) In *The Wiley Encyclopedia of Packaging Technology, 2nd ed.*, A.L. Brody and K.S. Marsh (Eds.), New York:John Wiley & Sons, pp. 335-343

7 Printing, Labeling, and Pigmenting

Printing or labeling on plastic packages is almost always essential to communicate important information to the purchaser or user of the product; to meet government requirements; and to stimulate sales. Plastics may be needed to be pigmented to protect products from light, but pigment is more often used to provide an attractive package to promote sales.

7.1 Printing

Printing methods for packaging, as well as for other purposes, can be divided into five major categories: relief, planographic, intaglio, porous, and impactless printing [1]. They can be further subdivided by whether the image is transferred directly from the image carrier to the item to be printed, called direct printing; or whether the image is first transferred to an intermediate, rubber-covered blanket cylinder, called offset printing [2]. The images can be created by chemical, mechanical, or electronic imaging methods. Photomechanical methods are used most often currently, but electronic imaging is growing rapidly.

Flexible packaging materials are generally printed while in webstock (roll) form, before conversion into the finished package shape. Molded containers must, by necessity, be printed after forming, but are generally printed before product is added. Thermoformed lids can sometimes be printed before forming. Even on flexible packages, however, some printed items, such as date and lot codes, are often added after the package is formed, filled, and sealed.

7.1.1 Relief Printing

Relief printing is also known as letterpress printing. In this method, the images, or printing areas, are raised above the surrounding, nonprinting areas. The ink rollers transfer ink to the top surfaces of the raised areas, which in turn transfer the ink to the package. The printing plates are usually made of rubber or, increasingly, of a photopolymer. The printing presses themselves can be platen, flatbed (or cylinder), or rotary. The ink travels through a series of rollers before being delivered to the plate to ensure that a precisely controlled amount of ink is deposited.

Flexography is a subcategory of relief printing in which the printing plates are flexible elastomers (rubber or plastic), and thin, highly fluid, rapid-drying inks are used. Flexographic printing was developed primarily for printing flexible packaging materials, and is widely used for this purpose. In fact, it is estimated that about 64% of all package printing is done by flexography, with offset and gravure used for most of the rest [3].

Relief printing, when magnified, can generally be recognized by the ghost-like ring of ink surrounding the image, caused by the pressure of the plate on the substrate which results in some spreading of the ink.

A variant of relief printing is hot stamping. In this process a web, or foil, with several layers of material, consisting of inks and coatings, between a sizing (a heat-activated adhesive) and a release-coated carrier material (most often PET), is used. A hot die, engraved with the image to be printed, presses the foil against the object to be printed. The die is most often made of silicone rubber or metal. The inks and sizing adhere to the container or other object in the die area. Hot stamping is often used to provide a mirror-like, metallic image on a plastic container [4].

Another variant of relief printing is the dry offset process, commonly used for containers. After the ink is distributed onto the raised surface of the printing plate, it is transferred to a rubber blanket. Images from as many as five additional plates are added and then the whole image is transferred to the container. A paste type of ink is used [5].

7.1.2 Planographic Printing

In planographic printing, or lithography, the printing image and the background are on the same plane of a thin metal plate. The plate has been treated to attract water and repel ink in the non-image areas, and do the reverse in the image areas. Both ink and water are transferred to the cylinder by a system of rollers. The image on the plate is transferred (offset) to another cylinder that is covered with a rubber blanket, and then finally to the substrate being printed. The primary advantage of this method is the clear sharp images that result.

7.1.3 Intaglio Printing

Intaglio printing is also known as gravure or rotogravure. This process can be considered the opposite of relief printing, in that the image to be printed is sunken below the level of the nonprinted background. Instead of using plates attached to the printing cylinder, as in the methods described earlier, the image is actually etched or engraved, either chemically or electromechanically, into the copper-plated cylinder in the form of tiny cells. Digital data can be used directly to engrave the cylinder by a diamond stylus, laser, or electron beam. The cylinder surface is then chrome-plated to resist wear.

The cylinder rotates in an ink bath, and the cells are filled with ink. Excess ink is wiped off the surface by a doctor blade, and then the image is transferred to the substrate as an elastomer-covered impression cylinder presses it against the printing cylinder. Gravure printing is excellent for long runs, since there is little wear on the cylinder, but preparation of the cylinders is very costly and they cannot be modified once they are prepared. They can, however, by recycled by removing the copper layer from the cylinder and then replating it.

The type of intaglio printing most often used for plastic containers is pad printing. In this process, the ink is deposited on an etched metal plate or cylinder. A doctor blade removes the ink from the non-etched portions and then the ink is picked up by a silicone rubber pad, which transfers the ink to the substrate. Thus, it is an indirect, or offset, type of printing. The pad is soft and flexible enough to conform to almost any container shape, and deposits all the ink it carries on the substrate. Colors are printed in sequence, and can be added while previous colors are still wet, because of the thinness of the deposit and the rapid drying characteristics of the ink [6]. The depth of the cells in gravure and pad printing is typically about 25.4 μm (0.001 in) [6].

7.1.4 Porous Printing

Porous printing is also known as screen printing. It involves forcing the ink through a stencil onto a substrate. The stencil may be made of silk, a synthetic fabric, or stainless steel, and can be prepared photographically. The mesh count of the screen can range from 30 to 200 per cm. Screen printing produces a much heavier layer of ink than the processes discussed earlier, and is ideally suited to printing on rounded and uneven surfaces such as bottles and tubes. It is especially suitable for small production runs because of the fairly low cost of the associated equipment [7]. Screen printing is the most common printing method for plastic bottles [3].

Inks can be cured by air-drying, ovens, or ultraviolet light; the last method is currently seeing the fastest growth.

7.1.5 Impactless Printing

Impactless printing is more commonly known as ink-jet printing. In this process, jets spray out electrically charged drops of ink, and the printer electrostatically directs these drops to the desired location on the substrate, forming characters. The excess drops fall into a reservoir and are reused. Ink jet printing is very widely used in packaging for adding date codes, lot numbers, and similar variable information to packages after they have been filled. Computer-driven systems that provide accurate, attractive, easily-changed images of all types on packages are taking over the market.

7.2 Labeling [8]

Labeling is most often used, instead of printing, to provide information on bottles and other containers, largely because it is much more complex and expensive to print shapes other than flat ones. Labels can be made of either plastic or paper, and can also incorporate foil for decoration. Paper labels often incorporate a polymer coating over the printing to protect from scuffing and moisture damage.

Labels can be affixed by wet glue, pressure-sensitive, or heat-sensitive adhesives. Solvent-based adhesives were widely used at one time, but have been largely replaced by water-borne and hot-melt adhesive systems. Labels are categorized as die cut or roll types. Die cut labels are dispensed from a stack of cut labels. Roll types, as the name indicates, are applied from a roll.

The labeling operation involves picking the label up from the stack or roll, applying or activating the adhesive if required, and, simultaneously or sequentially, placing it in position on the container. Coupled with this, in many applications, are systems that verify the label copy and its placement. The pharmaceutical industry, in particular, has very stringent requirements to ensure that products are correctly labeled. In the U.S., pharmaceutical labels must be roll-type rather than stack labels. Printing and labeling can be coupled so that variable information, such as date codes, is imprinted on the label either before or after it is applied.

One interesting recent development in labeling is use of transparent film as the label background. This label appears to blend into the container, resulting in a "printed look" to a labeled bottle. This works especially well if the bottle has been in-mold labeled. The film label is generally reverse-printed, so the image is protected from scuffing and maintains a fresh appearance. The film substrate must have excellent transparency, of course.

7.2.1 Wet-Glue Adhesive Labels

The most common labeling method is to apply a water-based glue to the back of the label, and then place the label on the object. This method is particularly effective with paper labels, and in very high speed applications. Most adhesives used today, with the exception of hot melts, are water-based, often in the form of a suspension containing polyvinyl acetate (PVA) or ethylene vinyl acetate (EVA) as the base polymer. The adhesive is applied to the label by a system of rollers that deliver a metered amount of adhesive. Pressure then affixes the label to the container or other item. Plastic labels are not generally suitable for use with wet-glue adhesives because they dry more slowly. Rapid drying is necessary for high productivity and for the label to stay in place once affixed.

Hot melt adhesives, which are 100% solid when cool and become liquid by heating, are suitable for use on plastic labels since they do not need to dry to form a good bond to the package.

7.2.2 Self-Adhesive Labels

As the name indicates, self-adhesive labels, or pressure-sensitive labels, are designed to adhere to the package without requiring activation by moisture, heat, or some other mechanism. The adhesive remains tacky and requires only light pressure for it to adhere to the substrate. The labels typically come in roll form on a silicone coated backing paper, are peeled off, and applied to the package. The label is released from the backing by bending the backing around a sharp angle. The label does not follow the bend immediately, freeing its leading edge, and allowing it to be picked up by a transfer mechanism. A variety of adhesives are used, although most are hot melts or water-based acrylics. The label substrate can be either paper or plastic.

The cost per label for pressure-sensitive labels is generally significantly higher than for glued labels. However, in many applications the reliability of the labeling operation and consequent decrease in downtime, along with lower maintenance requirements of pressure-sensitive application systems can more than make up for the higher cost. The equipment to apply pressure-sensitive labels is also less costly than glue-based systems. Pressure-sensitive labels have grown in use significantly as a result.

Self-adhesive labels can be designed to be either permanent or removable (repositionable). In most cases, permanent labels are used, but removable labels are needed in certain applications.

7.2.3 Heat-Sensitive Labels

Heat-sensitive labels, as their name implies, are activated by heat. They can be subdivided into two categories. Instantaneous heat-sensitive labels are heated to activate the adhesive when the label is in place on the package, sealing the label to the package. Delayed action heat-sensitive labels are heated before being positioned, rendering them pressure-sensitive, and are then applied to the package. Delayed action heat-sensitive labels are most often used on plastic containers, with the exception of in-mold labels. In-mold labels, discussed in Section 5.3.5, are the instantaneous heat-sensitive type.

7.2.4 Heat Transfer Labeling

Heat transfer labeling can be regarded as a cross between labeling and printing. The label copy is reverse-printed onto a special substrate by conventional printing techniques. The label is then placed in contact with the surface, print side against the container, and heat is applied. The heat causes the inks to bond the image on the label to the object being printed. The label substrate is then peeled away, leaving the image on the container surface.

7.2.5 Shrink or Stretch Sleeves and Bands [9]

Another option for affixing labels to containers is to use a shrink or stretch sleeve or band without an adhesive. Shrink sleeves are commonly made of monoaxially oriented PVC, PP, or PETG. Shrink sleeves are larger than the container and are shrunk 50 to 70% to fit using heat. Stretch sleeves are most often made of LDPE, and are stretched to fit in place around the container. In both cases labels are held in place by the elasticity of the film.

Both shrink and stretch sleeves are generally printed on the inside surface in flat webstock form, by rotogravure or less commonly by flexography. The webstock is often assembled into tube form using solvent seaming, to avoid the distortion which heat sealing can cause. In both stretch and shrink sleeves, the printed image will be distorted by the change in the film dimensions. Therefore the printed image must be distorted so that the final image, after stetching or shrinking, will be distortion-free. Uniaxially oriented film makes this easier to achieve than biaxially oriented film, and also minimizes wrinkling.

An alternative to the use of seamed webstock is the use of tubing, extruded in the desired width and then printed. Tubing is generally less expensive, but it cannot be reverse printed.

Shrink bands are used to provide tamper-indicating features on bottles and jars. These heat-shrink labels are typically made with a relatively brittle PVC film and perforated so that they are fractured on opening. U.S. government regulations require that the bands are printed when used for mandatory tamper-evident features, so they cannot be readily replaced by look-alike bands. When the shrink band also is the primary product label, the tamper-evident feature can be separated from the body label by horizontal perforations, permitting removal of part of the band for access to the product.

Another common use of shrink bands is to bundle items together for retail sale. The bundle may include multiples of the same product, combinations of different but related products, or a promotional item and a product.

Expanded PS shrink band/labels are used on glass bottles to prevent damage during distribution and enable the bottles to be downguaged. Glass is very sensitive to surface damage, so the cushioning properties of the expanded PS are significant in this application.

7.3 Pigmenting

Color is often an important attribute of a plastic package. It can be just as much of an identification and marketing tool as is the printing or labeling of the container. Color may also be necessary to enhance a plastic's light barrier for product protection. The color may be obtained by printing the whole package, but is much more likely to result from pigmenting the plastic before it is formed into the package shape.

Dyes can also be used, but are much less common than pigments. The difference between the two is that dyes dissolve in the material they are coloring and produce color by selective absorption of light without scattering. Pigments remain solid and produce color by selective absorption as well, but also scatter light. As a result, dyes do not affect the transparency of the material they are coloring. Pigments generally render the material opaque, although some pigments with a very small particle size scatter very little light and therefore have little effect on transparency [10].

Pigments for polymers are generally in the form of a master batch, in which a concentrated dose of the pigment is blended with a relatively small amount of carrier resin. When the packaging material is produced, a metered amount of this master batch is fed into the extruder along with the base resin. Within the extruder, the master batch is blended with the base resin, and the pigment (and other additives) evenly dispersed. Ideally, the carrier resin in the master batch should be the same as the base resin. However, some master batch systems are available which are suited to a range of base polymers.

Two of the most widely used pigments for plastic are carbon black, to achieve a black color, and titanium dioxide, to produce a white color. A variety of other materials are used to produce a wide range of colors. Coloring agents for food packages, of course, must meet requirements for food contact applications.

7.4   Heavy Metals in Inks and Pigments

In the 1980s, several U.S. states, especially on the east coast, began to rely more on incineration of municipal solid waste as an alternative to disposal in landfills. As a result, incinerator ash was generated. When the ash was tested, it often contained unacceptable levels of heavy metals, which meant that its disposal had to be done in accordance with U.S. hazardous waste regulations, rather than simply as municipal solid waste. For several years there were regulatory questions about whether the U.S. Congress had intended incinerator ash to fall under hazardous waste regulations or be exempt. The eventual ruling was that this ash was not exempt. Diposing of this ash according to hazaardous waste regulations was much more expensive to municipalities because hazardous waste landfills are more expensive to build and operate than municipal solid waste landfills. Thus, state governments had an incentive to reduce the toxicity, in particular the heavy metal content, of their incinerator ash. The specific heavy metals of concern were lead, cadmium, mercury, and hexavalent chromium.

Most of the incineration facilities were of the mass-burn type, where essentially unsorted garbage is combusted. The largest source of heavy metals in municipal solid waste is automotive batteries, a major source of lead. Several states placed deposits on automotive batteries to remove them from the waste stream. After batteries, packaging materials, especially the inks and plastic pigments they contain, and to a lesser extent plastic stabilizers, were a significant source of heavy metals. Several states in the northeastern part of the U.S., under the auspices of the Coalition of Northeastern Governors (CONEG), established a Packaging Task Force to look at this and other issues related to solid waste. The task force included representatives of industry and environmental groups, as well as representatives of the states' legislative bodies.

As a result, CONEG drafted a model state law known as CONEG's Model Toxics Legislation, which called for the elimination of the target heavy metals - lead, cadmium, mercury and hexavalent chromium - in packaging applications. The precise wording of the legislation was that deliberate introduction of these heavy metals during manufacture or distribution of packaging was prohibited. Incidental introduction of the metals was to be decreased according to the following schedule: no more than 600 ppm total within two years of the signing of the legislation into law; no more than 250 ppm within three years after it became law; and no more than 100 ppm at the four year mark. Intentional introduction is "... the act of deliberately utilizing a regulated metal in the formation of a package or packaging component where its continued presence is desired in the final package or packaging component to provide a specific characteristic, appearance or quality." [9] Incidental introduction is defined as circumstances where the regulated metal is an unintended or undesired ingredient, such as in the case of a processing agent or catalyst, where the residue is not intended to be present [9].

The legislation provides for exemptions for packages or components manufactured before the effective date of the legislation; packages where the guidelines are exceeded solely as a result of the use of recycled materials; packages where the metals are used to comply with federal health or safety requirements and an exemption petition has been accepted; and packages for which there is no feasible alternative and an exemption has been granted. It is specified that "no feasible alternative" applies only when the metal is essential to the protection, safe handling, or function of the package contents. Exemptions are granted only for two years, and re-applications must be submitted for exemptions to be extended. The legislation applies to all packaging used or sold within the state adopting the legislation. There is no exemption for imported goods.

In addition to the reduction in toxicity of incinerator ash, it was also considered desirable to eliminate heavy metals from landfills to prevent their seepage into groundwater aquifers. Thus, even states which do not incinerate much of their municipal solid waste have had an interest in this legislation. A list of states that have adopted the CONEG Model Toxics Legislation as of June 1997 is presented in Table 7.1. Many of these states have slightly modified the model legislation.

Table 7.1 States Adopting CONEG's Model Toxics Legislation* [11]

Connecticut	Florida	Georgia
Illinois	Iowa	Maine
Maryland	Minnesota	Missouri
New Hampshire	New Jersey	New York
Pennsylvania	Rhode Island	Vermont
Virginia	Washington	Wisconsin

* as of June, 1997

As can be seen, the states adopting the Model Toxics Legislation are some of the most populous in the U.S. and account for a substantial part of the market. Therefore, even though this legislation is not in effect in all states, it has become a *de facto* U.S. standard, and has led to some significant packaging changes.

Until the passage of this legislation, it was not uncommon for inks and pigments for plastics and other materials to be manufactured using cadmium and hexavalent chromium, to achieve some desired colors. Lead stabilizers were used in some PVC packaging, although not in food packaging, and were also used in some inks and pigments.

These materials have now, for the most part, disappeared from inks and pigments for packaging uses. The changes to alternative inks and pigments, often organic, usually resulted in additional cost, and sometimes in some sacrifice in package appearance, particularly for graphics.

7.5 Surface Treatment

On some plastics, especially polyethylene, inks and to a lesser extent, adhesives and coatings, tend not to adhere well. For good quality printing, surface treatment is generally necessary. Surface treatment produces polar groups on the surface of the polymer, which increase its ability to bond to inks and adhesives. This oxidative process can also remove from the surface contaminants which may interfere with adhesion.

Three major techniques are used. Corona discharge is most often used for plastic film or sheet. In this process, in the presence of heat and air at atmospheric pressure, the plastic moves between an electrically grounded roller and a high voltage electrode. The air between the two surfaces becomes ionized, resulting in a continuous arc discharge at the surface of the film or sheet. This discharge cleans and oxidizes the surface. Ozone is generated as the active oxidizing agent, and must be removed from the workplace environment, generally by catalytic means. It is also important to maintain low relative humidity to prevent excessive chain scission, which can result in a frosted appearance of the surface. The effectiveness of corona discharge generally decreases with time, sometimes quite rapidly [12,13].

Gas plasma techniques are most commonly used with containers. The containers are placed in a chamber which is evacuated and then charged with oxygen, argon, helium, or nitrogen. A radio frequency field ionizes the gas, producing a glow discharge which oxidizes the surface when oxygen is the gas used. The other gases can activate the surface also. This process is done at relatively low temperatures and pressures [12].

The most common technique for treating containers is flaming, where the containers pass by a bank of flame jets, usually fueled by natural gas. The flame oxidizes the surface and burns off surface contaminants.

Less commonly, polyolefins can be treated with fluorine or sulfur trioxide, producing a thin polar surface with enhanced adhesion for printing and adhesives. Some cross-linking of the surface occurs, which renders heat sealing more difficult. Fluorination and sulfonation are also important for barrier improvement, and are discussed further in Sections 5.5.2 and 5.5.3.

Another common surface treatment is coating, which is commonly used to provide a base coat for the printing or to provide an overcoat on the printed surface to protect it from damage. Titanium dioxide, which provides a white color for printing, is a common base coat [2].

7.6 Reverse Printing

Reverse printing is unique to transparent films, whether they are used in packages or in labels. In this method, the printing is done on the back of the film, which is often then laminated to another layer of material. The printing is then seen through the clear plastic film, and the film protects the printing against scuffing, abrasion, and other damage.

Closely related is the practice of printing labels for transparent bottles on both sides, so that the reverse side of the label can be viewed through the product and the other side of the bottle. Obviously this technique is most applicable when the product itself, as well as the container, is extremely clear.

One interesting application of this double-sided labeling is the Soft Soap (Colgate-Palmolive Co.) containers currently on the market. The containers have an internal label as well as external labels on the front and back. The presence of the internal label plus the ability to see the back label through the front of the container (and the clear product) provides a very attractive, three-dimensional, underwater appearance to the labeled bottle.

References

1. Lentz, J. (1986) In *The Wiley Encyclopedia of Packaging Technology*, M. Bakker (Ed.), New York: John Wiley & Sons, pp. 554-559
2. Taggi, A.J. and Walker, P.A. (1997) In *The Wiley Encyclopedia of Packaging Technology, 2nd ed.*., A.L. Brody and K.S. Marsh (Eds.), New York:John Wiley & Sons, pp. 783-787
3. Eldred, N.S. (1993) *Package Printing*, Plainview, NY:Jelmar Pub. Co., Inc.
4. Brody, A.L. and Marsh, K.S. (1997) In *The Wiley Encyclopedia of Packaging Technology, 2nd ed.*, A.L. Brody and K.S. Marsh (Eds.), New York:John Wiley & Sons, pp. 296-297
5. Brody, A.L. and Marsh, K.S. (1997) In *The Wiley Encyclopedia of Packaging Technology, 2nd ed.*, A.L. Brody and K.S. Marsh (Eds.), New York:John Wiley & Sons, pp. 298-300
6. Brody, A.L. and Marsh, K.S. (1997) In *The Wiley Encyclopedia of Packaging Technology, 2nd ed.*, A.L. Brody and K.S. Marsh (Eds.), New York:John Wiley & Sons, pp. 300-302
7. Brody, A.L. and Marsh, K.S. (1997) In *The Wiley Encyclopedia of Packaging Technology, 2nd ed.*, A.L. Brody and K.S. Marsh (Eds.), New York:John Wiley & Sons, pp. 302-303
8. Brody, A.L. and Marsh, K.S. (1997) In *The Wiley Encyclopedia of Packaging Technology, 2nd ed.*, A.L. Brody and K.S. Marsh (Eds.), New York:John Wiley & Sons, pp. 536-541
9. Kopsilk, D.F. and Fisher, E. (1997) In *The Wiley Encyclopedia of Packaging Technology, 2nd ed.*, A.L. Brody and K.S. Marsh (Eds.), New York:John Wiley & Sons, pp. 69-70
10. Brody, A.L. and Marsh, K.S. (1997) In *The Wiley Encyclopedia of Packaging Technology, 2nd ed.*, A.L. Brody and K.S. Marsh (Eds.), New York:John Wiley & Sons, pp. 242-256
11. Thompson Publishing Group (1997) *Environmental Packaging: U.S. Guide to Green Labeling, Packaging and Recycling*, Washington, D.C.
12. Brody, A.L. and Marsh, K.S. (1997) In *The Wiley Encyclopedia of Packaging Technology, 2nd ed.*, A.L. Brody and K.S. Marsh (Eds.), New York:John Wiley & Sons, pp. 864-867
13. Finson, E. and Kaplan, S.L. (1997) In *The Wiley Encyclopedia of Packaging Technology, 2nd ed.*, A.L. Brody and K.S. Marsh (Eds.), New York:John Wiley & Sons, pp. 867-874

8 Barrier, Migration, and Compatibility

8.1 Introduction

The shelf life of a product, i.e., the length of time it remains in an acceptable condition for sale, is often determined by the ability of its package to prevent loss of desirable constituents from the product and to prevent gain of substances adversely affecting product quality. These losses or gains may be to or from the surrounding environment or the package itself. Thus, for example, a soft drink can "go flat" by losing carbon dioxide through the container wall to the atmosphere. A fruit-flavored cereal can lose its flavor as a result of "scalping" by its plastic pouch. Potato chips can become rancid because oxygen passes through the package and reacts with the product. The first PVC liquor bottles were taken off the market because of concerns about migration of residual vinyl chloride monomer from the bottle to the product. Usually, but not always, the substances gained or lost are either gases or are volatile.

While these concerns exist to some degree for other packaging materials, they are most often associated with plastics. The reason for this is simple. Glass, steel, and aluminum are in essence perfect barriers to the transport of materials. The compatibility of these packaging materials with the products they contain is determined by their chemical composition, no lead glass or soldered steel cans, for example. Another way of guaranteeing compatibility is by incorporating barriers, often plastics.

Plastics, with their variety of chemical compositions and performance as well as other advantages, can substitute for glass and metal in many packaging applications, but only if we understand their limitations. Plastics are, in general, permeable to various substances. When there is a difference in a substance's chemical activity, usually expressed in terms of concentration, between one side of the plastic and the other, the result is a net movement of the substance (permeant) from the high concentration side to the low concentration side, which continues until the chemical activities are equal on both sides.

8.2 Barrier

The barrier properties of a plastic refers to its abilities to retard the transfer of something through the plastic. Usually the "something" we refer to is a chemical substance, but plastics can also provide a barrier to visible light, ultraviolet light, or heat. Most plastics are poor barriers to light unless pigmented to render them opaque. Plastics generally have fairly low coefficients of thermal conductivity, so are fairly good heat barriers, especially plastic foams, which are often used as insulators. In this chapter, most of the discussion concerns plastics as barriers to chemical substances.

When a polymer is exposed to a high concentration of permeant on one side and a lower concentration on the other side, and the permeant passes through the polymer, several steps occur, as illustrated in Fig. 8.1. First, the permeant molecule dissolves in the polymer; then it diffuses through the polymer, and finally it desorbs from the polymer on the other side. Actually the transfer occurs in both directions, but the net effect is transfer from the high concentration side to the low concentration side. How rapidly this process occurs is a function of both how readily the permeant dissolves in the polymer (its solubility constant) and how fast the permeant molecules move within the polymer matrix (its diffusivity).

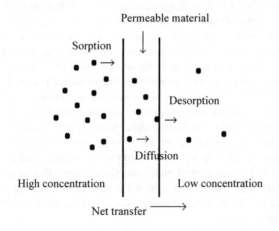

Figure 8.1 Permeation of a substance through a plastic packaging material

8.2.1 Permeability Constant

The barrier capability of a plastic film is usually expressed in terms of its permeability constant, $\overline{P}$. It represents the rate at which a quantity of permeant passes through a unit surface area in unit time, depending on the film thickness, temperature, and the concentration (partial pressure) difference for the permeant on the two sides of the plastic film. At a given temperature, the equation that relates the permeability constant to the steady state mass transfer through the polymer is:

$$\overline{P} = \frac{Q\,L}{A\,t\,\Delta p} \tag{8.1}$$

where

$\overline{P}$ is the permeability coefficient
Q is the mass of permeant passing through the material
L is the material thickness
A is the surface area available for mass transfer
t is the time
Δp is the partial pressure difference for the permeant between the two sides of the material

When permeation occurs largely at steady state; storage conditions are constant; the length of time involved is relatively long so that transient effects associated with initial conditions (and possibly with openings and reclosings) are not significant; and Δp remains essentially constant during the storage period; this equation can be used easily to estimate shelf life. All that is required is knowledge about the sensitivity of the product to gain or loss of the permeant involved and the package and storage characteristics. Similarly, once the product characteristics are known and the desired shelf life has been determined, the permeability equation can be used to screen potential package structures, determining which are viable candidates for a more thorough analysis taking into account cost, availability, and other factors. Of course in the real world, things are seldom this simple. Nonetheless, this equation can provide a starting point for package design.

It should be noted that the permeability constant of a packaging material and the permeability constant of a package manufactured from that material are not necessarily identical. The process of forming a package may change material characteristics as a result of heat history and associated changes in crystallinity, changes in degree of orientation, the introduction of microvoids, leading to increased permeation, or other factors. Actual pinholes or leaks in the package structure can result in very large increases in mass transfer which are not accommodated in this model.

Differences in the thickness of various parts of the container induced by the forming process can make it difficult to apply this simple permeability equation. For rigid containers, the influence of the closure system can be significant, both in terms of microleaks and of differences between the barrier properties of the closure and liner and those of the container. Thus, calculations based on permeability characteristics of materials are only estimates of the behavior of actual package systems. The simpler the model used, the more it is likely to deviate from package behavior in the environment. On the other hand, even simple models can help determine the best plastic packaging materials for the application at hand.

One such simple model says that the permeability constants of a polymer for nitrogen, oxygen, and carbon dioxide are generally in the ratio of 1:4:14 respectively [1]. Barrier for water vapor does not correlate well with barrier for these gases.

8.2.2 Relationship Between Permeability, Diffusivity, and Solubility

One of the first refinements in using the concept of permeability is recognizing that, while Eq. (8.1) represents a reasonably accurate representation for steady state mass transfer through packages, in a great many cases the transfer is not at steady state. When crackers absorb moisture and lose their crispness, for example, the water vapor partial pressure inside the package is changing over time, even if the surroundings remain constant (which is often not the case, either). To understand when Eq. (8.1) is applicable, and the errors that may arise when using it, a more fundamental analysis is necessary.

The equation describing mass transfer across a permeable barrier is :

$$\frac{dQ}{dt} = \frac{1}{L} \bar{P} A (p_2 - p_1) \tag{8.2}$$

where p_2 and p_1 represent the permeant partial pressure on the two sides of the polymer.
    This equation in turn is based in Fick's laws of diffusion

$$F = -D \frac{dC}{dx} \tag{8.3}$$

where F is the flux of permeant, D is the diffusion coefficient, C is concentration, and x is position within the film (from 0 to L); and

$$\frac{\partial C}{\partial t} = D \frac{\partial^2 C}{\partial x^2} \tag{8.4}$$

combined with Henry's Law representing the relationship between solubility and partial pressure (accurate for ideally dilute solutions):

$$C = S\ p \qquad (8.5)$$

where S is the Henry's law constant (solubility constant) for the permeant/polymer combination and p is partial pressure of the permeant.

At steady state, by definition, the concentration does not change with time, so Eq. (8.4) is equal to zero. If the diffusion coefficient D is constant, Eq. (8.4) can be integrated twice with respect to x, and the boundary conditions shown in Fig. 8.2 can be applied, resulting in the relationship

$$\frac{C - C_1}{C_2 - C_1} = \frac{x}{L} \qquad (8.6)$$

for any position x in the plastic package.

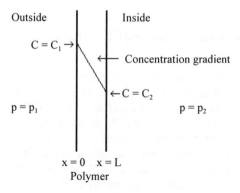

Figure 8.2   Steady state diffusion

Substituting Eq. (8.6) in Eq. (8.3) and integrating with respect to x, we obtain

$$F = D\ \frac{C_1 - C_2}{L} \qquad (8.7)$$

Henry's law then tells us that $C = S \, p$, so:

$$F = D \, S \, \frac{p_1 - p_2}{L} = \bar{P} \, \frac{p_1 - p_2}{L} \tag{8.8}$$

where $\bar{P}$ is defined as the diffusion coefficient D times the solubility constant S. Knowing that the flux, F, is:

$$F = \frac{Q}{t \, A} \tag{8.9}$$

and rearranging, we then finally reach Eq. (8.1).

It is Eq. (8.1) that has led to the common systems of units for permeability constants, where values are given in:

$$\frac{\text{mass thickness}}{\text{area time (partial pressure difference)}}$$

8.2.3   Applications of Permeability Equations

We can see that in deriving the permeability equation (Eq. 8.1), a number of crucial assumptions were made, which are not true under all conditions. One of the most obvious assumptions is that the concentrations, and therefore partial pressures, of the permeant inside and outside the package are constant. While this is often a reasonable approximation for oxygen-sensitive products, as discussed earlier, it is obviously not accurate in many packaging situations; a major example is the loss of carbon dioxide from a PET soft drink bottle.

In this case, we could start with Eq. (8.2), and insert a term describing the relationship between t and p_2 (the CO_2 partial pressure inside of the bottle) before integrating. A similar case, widely encountered in selecting packaging materials for water vapor barrier, is when the partial pressure of water vapor inside the package changes during storage, even if the outside conditions remain constant.

Another assumption in Eq. (8.1) is that Henry's Law accurately represents the relationship between solubility and partial pressure. This is generally valid for polymers at temperatures above their glass transition temperatures and when the partial pressures of penetrants are low. For glassy polymers and higher partial pressures, a nonlinear Langmuir-Henry's law model is often followed [2]:

$$C = kp + \frac{c_H\, bp}{1 + bp} \qquad\qquad (8.10)$$

where k is the Henry's law constant; c_H is the Langmuir capacity constant; and b is the Langmuir affinity constant. In this case, $\overline{P}$ is a function of the partial pressure, so this relationship must be inserted before integrating Eq. (8.2).

A third important assumption is that the diffusion coefficient is independent of permeant concentration. For organic vapors, this is often not the case except at very low concentrations. Further, the diffusion and solubility coefficients can vary during storage in response to other influences. It is well known, for example, that for water-sensitive polymers such as EVOH, moisture can significantly affect permeability coefficients, largely through its influence on diffusivity. In these cases, it is necessary to incorporate in the governing equation the dependence of the terms on not only the concentrations of the diffusing species of interest (e.g., oxygen), but also on the concentrations of copermeants such as water vapor.

Further, the influence of varying temperatures has not been considered in the above analysis. Both solubility and diffusion coefficients generally are affected by temperature, with D increasing with increasing temperature, following Arrhenius' law:

$$D = D_o e^{(-E_a/RT)} \qquad\qquad (8.11)$$

where E_a is the activation energy; R is the gas constant; and T is absolute temperature. The solubility, on the other hand, generally decreases with increasing temperature for gases, while increasing for liquids and solids.

As can be seen, the basic permeability equation (Eq. 8.2) can easily become too complex to solve analytically. In such cases, numerical methods for integration can be employed to permit shelf life calculations. Far more frequently however, simplifying assumptions are made, with the understanding that the resultant shelf life estimations are less accurate.

8.2.4 Oxygen Transmission

The shelf life of many products is affected by oxidation associated with the entry of oxygen into the package, as well as reaction with oxygen originally contained inside the package, either in the headspace or within the product. Oil-containing foods become rancid; vitamin C oxidizes and is lost; various types of food undergo color changes, among other effects. The kinetics of the product's reaction with oxygen are often highly complex and poorly understood, rendering it very difficult to accurately describe the relationships between the amount of oxygen permeating into the package; the oxygen concentration inside the package; the oxygen concentration within the product; and the extent of the degradative reactions. Typically, estimations of shelf life are based on specifing the amount of oxygen which the product can absorb before reaching the end of its shelf life condition. Further, it is typically assumed that residual oxygen inside the package is quickly consumed by reaction; steady state permeation of oxygen is reached relatively quickly; and any oxygen permeating into the package is relatively quickly consumed, so that the oxygen partial pressure inside the package rapidly becomes, and remains, very close to zero. Thus Eq. (8.1) can be applied. The partial pressure difference, Δp, is 0.21 atm (partial pressure of oxygen in the earth's atmosphere), since p inside the package is assumed to be zero. While for many products this is not a very good assumption, it is often the best that can be made in the absence of thorough understanding of the reaction kinetics involved. The following examples, illustrating both metric and English units, use Eq. (8.1) to determine the shelf life of common products:

Example 1:

It has been determined that for a 100 g packet of peanuts packaged in PVDC film, the allowable oxygen gain is 15 ppm. The thickness of the film is 25 μm. The surface area is 200 cm². $\overline{P}$ = 30 cm³ μm/m² d atm at the (constant) storage temperature. What is the shelf life of the peanuts, assuming there is no oxygen initially inside the package?

Answer 1:

Q = 0.000015(100 g)(mol/32 g)(22,400 cm³/mol) = 1.05 cm³
t = (1.05 cm³)(25 μm)(30 cm³ μm/m² d atm)$^{-1}$(200 cm²)$^{-1}$(m²/ 10000 cm²)$^{-1}$(0.21 atm)$^{-1}$ = 208 days

Example 2:

It has been determined that for 2 lb of peanut butter packaged in a PET jar, the allowable oxygen gain is 100 ppm. The average thickness of the jar is 8 mil (0.008 in). The surface area is 100 in². $\overline{P}$ = 5 cc mil/100 in² 24 h atm at the (constant) storage temperature. What is the shelf life of the peanut butter, assuming there is no oxygen initially inside the package?

Answer 2:

Q = .0001(2 lb)(454 g/lb)(mol/32 g)(22,400 cm³/mol) = 64 cm³
t = (64 cc)(8 mil)(5 cc mil/100 in² 24 h atm)$^{-1}$(100 in²)$^{-1}$(0.21 atm)$^{-1}$(d/24h) = 490 days

For highly oxygen-sensitive products, such as peanuts, the amount of oxygen contained in the headspace of the package or entrained in the product can be enough to cause significant deterioration, even if no additional oxygen permeates through the package. Therefore, for such products to have an acceptable shelf life, it is important to remove as much residual oxygen as possible from inside the package before it is sealed. Nitrogen flushing and vacuum packaging are commonly used for this purpose, although they are often not as effective as they are perceived to be in accomplishing oxygen removal. The problem can be compounded by failure to understand the dynamics of the gas flow process. For example, a frequent response when residual oxygen content in the package is found to be too high after gas flushing is to increase the flow rate of nitrogen. However, to get good removal of oxygen from the package, a laminar flow of gases is needed. Increasing the flow rate is likely to transform what may have been laminar flow into turbulent flow, resulting in pockets of air in the package which are not affected by the flow at all. In some cases, a reverse flow of air into the pouch as the turbulent nitrogen stream exits may result, thus increasing the level of residual oxygen in the package and making a bad situation worse instead of better. Similarly, vacuum packaging does not produce a total vacuum in the bag, and the vacuum is not maintained at its original level unless the packaging material is totally impermeable.

8.2.5 Water Vapor Transmission

For products sensitive to the gain or loss of moisture, the water vapor concentrations both outside and inside the package are generally far from zero, and the simplifying assumption used for oxygen leads to major errors in shelf life estimation. Further, the relationship between the moisture content of products and the water vapor partial pressure in the atmosphere surrounding those products at equilibrium is generally nonlinear.

A representative moisture sorption isotherm is shown in Fig. 8.3. Fortunately, the water concentrations of interest, representing acceptable product conditions, often fall on a fairly linear portion of the curve, which can be represented by:

$$W_i = a + bM \qquad (8.12)$$

where M is the percent moisture in the product on a dry weight basis; W_i is the water activity of the environment in equilibrium with product of this moisture content; and a and b represent the intercept and slope, respectively, of the best-fit line to the linear portion of the isotherm.

We generally assume that equilibrium is established quickly relative to the rate of permeation. We can also generally assume with reasonable accuracy that the amount of permeating water vapor remaining in the package headspace is negligible compared to the amount absorbed by the product (see Part 2 of Example 3). In that case:

$$Q = (M - M_i)w \qquad (8.13)$$

where w is the dry product weight and M_i is the initial moisture content, so:

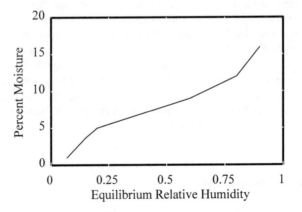

Figure 8.3 A moisture sorption isotherm for a product.

$$\frac{dQ}{dt} = w\frac{dM}{dt}$$ (8.14)

It is useful to rewrite the basic permeability equation (Eq. 8.2) in terms of water activity rather than concentration:

$$w\frac{dM}{dt} = \frac{\overline{P}Ap_s}{L}(W_2 - W_1)$$ (8.15)

where p_s is saturation vapor pressure and W represents water activity ($p = p_s W$). We can then substitute the relationship between W_1 and M (Eq. 8.12) to obtain:

$$\frac{dM}{dt} = \frac{\overline{P}Ap_s}{Lw}(W_2 - a - bM)$$ (8.16)

This expression can now be integrated:

$$\int \frac{dM}{W_2 - a - bM} = \frac{\overline{P}Ap_s}{Lw}\int dt$$ (8.17)

to give:

$$-\frac{1}{b} \ln \frac{[W_2 - W_1]_{t=t}}{[W_2 - W_1]_{t=0}} = \frac{\overline{P}Ap_s}{Lw} t \qquad (8.18)$$

If the form of the isotherm does not permit an analytical solution to the permeability equation, numerical methods can be used. Example 3 illustrates a shelf life calculation for a product with a linear isotherm.

Example 3, Part 1:

200 g (0.44 lb) of product is packaged in 50.8 μm (2 mil) LDPE film, with 1290 cm² (200 in²) total surface area and headspace of 10 cm³ (0.61 in³). The product is stored at 70% relative humidity (RH), 21.1 °C (70 °F). The product is packaged at 3% moisture, dry weight basis, corresponding to 20% RH, according to the product's moisture sorption isotherm. The highest acceptable moisture content is 8%, determined by organoleptic (taste) testing, corresponding to 67% RH, according to the moisture sorption isotherm. The sorption isotherm is observed to be close to linear in the region in question. It is assumed that the seals are intact, with no leakage through the seals or through other package defects. The permeability constant for water vapor for this LDPE film is 13.4 g μm/m² d mm Hg (0.034 g mil/100 in² 24 h mm Hg). Saturation vapor pressure at 21.1 °C (70 °F) is 18.77 mm Hg. What is the shelf life?

Answer 3, Part 1:

Note water activity W = RH/100
$b = (.67-.20)/(.08-.03) = 9.4$

$t = - (50.8 \text{ μm})(200 \text{ g})(13.4 \text{ g μm/m}^2 \text{ d mm Hg})^{-1}(1290 \text{ cm}^2)^{-1}(10,000 \text{ cm}^2/\text{m}^2)(18.77 \text{ mm Hg})^{-1}$
$(9.4)^{-1} \ln[(.70-.67)/(.70-.20)] = 94 \text{ days}$
$\{t = - (2 \text{ mil})(200 \text{ g})(0.034 \text{ g mil}/100 \text{ in}^2 \text{ 24 h mm Hg})^{-1}(200 \text{ in}^2)^{-1}(18.77 \text{ mm Hg})^{-1}(9.4)^{-1}$
$\ln[(.70-.67)/(.70-.20)](d/24 \text{ h}) = 94 \text{ days}\}$

Example 3, Part 2:

Check the assumption that the amount of water vapor in the headspace is negligible compared to the amount gained by the product.

Answer 3, Part 2:

The product gains 5% moisture: $0.05(200 \text{ g}) = 10 \text{ g}$
Moisture in the headspace increases from 20% RH to 67% RH, a gain of 47% RH. Water gain in the headspace: $0.47(18.77 \text{ mm Hg})(10 \text{ cm}^3)(82.06 \text{ cm}^3 \text{ atm mol}^{-1} \text{ K}^{-1})^{-1}(273 \text{ K})^{-1}(\text{atm} / 760 \text{ mm Hg})(18\text{g/mol}) = 5.2 \times 10^{-6} \text{ g}$.
The amount of water in the headspace is negligible compared to the amount gained by the product.

8.2.6 Water Vapor Transmission Rates

While the barrier capability of plastic materials is generally expressed in terms of permeability constants for oxygen, carbon dioxide, and organic vapors, tabulations of moisture barrier often list water vapor transmission rates instead. These are not true permeability constants, since they are a function of the water vapor pressure differential during the testing, as well as of the temperature. Generally these rates have been measured under a standard set of conditions, often determined by a national or international standards-setting body. The relationship between the water vapor transmission rate (WVTR) and the permeability constant for water vapor is as follows:

$$\overline{P} = \frac{WVTR}{(p_2 - p_1)_{test}} \tag{8.19}$$

where the partial pressure difference for water vapor, $(p_2 - p_1)$, is that used during the measurement of the water vapor transmission rate. Many tabulations use ASTM standard conditions, where the temperature is 35 °C (95 °F); the relative humidity on one side of the film is 90%; and there is a desiccant on the other side of the film to maintain the relative humidity at approximately zero. In these cases, WVTRs can be transformed to permeability constants by dividing by 0.90 times the saturation vapor pressure of water at that temperature, 42.175 mm Hg, as in the following example:

Example 4:

The WVTR for LDPE is 510 g µm/m² d (1.3 g mil/100 in² 24 h) as determined under standard ASTM conditions. What is the permeability constant?

Answer 4:

$\overline{P}$ = (510 g µm/m² d)/37.960 mm Hg = 13.4 g µm/m² d mm Hg
{$\overline{P}$ = (1.3 g mil/100 in² 24 h)/37.960 mm Hg = 0.034 g mil/100 in² 24 h mm Hg}

8.2.7 Units

Some rather strange sets of units are used for water vapor transmission rates and permeability constants, especially under the English system. The units used in the examples above, g mil/(100 in² 24 hr) for tabulations of WVTR, and cc mil/(100 in² 24 hr atm) for tabulations of $\overline{P}$, are those most commonly used in the U.S. In other parts of the world, various systems of units are used, often with a 20 or 25 μ film as the standard film thickness. Thus, care must be taken in using permeability constants and water vapor transmission rates, to ensure proper units are utilized, and unit conversions are carried out correctly. A tabulation of conversion factors for a number of different systems of permeability units and water vapor transmission rates is found in the Wiley Encyclopedia of Packaging Technology [1].

8.2.8 Effect of Temperature on Permeability Constants

As discussed earlier, diffusion rates generally increase with temperature, while the solubility tends to decline for gases and increase for solids and liquids. However, at least for relatively small temperature ranges, the relationship between permeability constants at different temperatures can be expressed by an Arrhenius type relationship:

$$\overline{P}_2 = \overline{P}_1 e^{(E_d/R)[(1/T_1)-(1/T_2)]}$$

(8.20)

The temperatures of interest must be both above or both below the polymer's glass transition temperature, as permeability increases significantly above T_g. Eq. (8.20) does not yield accurate results if one temperature is above T_g and the other below. Plots of permeability constants as a function of temperature that are linear for substantial temperature ranges show a discontinuous change at T_g, as activated diffusion assumes an important role in the transport process.

Activated diffusion refers to the contribution made to the diffusion process by the segmental mobility of polymers above their glass transition temperature. Below this temperature, there is no segmental mobility. In essence, the motion of the polymer opens up spaces in the polymer matrix as the diffusing molecule moves through, significantly increasing the diffusion rate.

The following example illustrates use of Eq. (8.20) to calculate the permeability constant for a polymer at one temperature given an activation energy and the value at another temperature:

Example 5:

The permeability constant for oxygen for PET is 0.12 x 10^4 cm^3 μm/m^2 day atm (3.0 cm^3 mil/100 in^2 24 hr atm) at 25 °C (77 °F). The activation energy for permeation is 43.8 kJ/mole. What is the permeability constant at 35 °C?

Answer 5:

$\overline{P}_{35}$ = 0.12 x 10^4 cm^3 μm/m^2 day atm exp[(43,800 J mol^{-1}/8.314 J mol^{-1} K^{-1})(1/298.16K - 1/308.16K)] = 0.21 x 10^4 cm^3 μm/m^2 day atm

$\{\overline{P}_{35} = 3.0 \text{ cm}^3 \text{ mil}/100 \text{ in}^2 \text{ 24 hr atm } \exp[(43,800 \text{ J mol}^{-1}/8.314 \text{ J mol}^{-1} \text{ K}^{-1})(1/298.16K - 1/308.16K)] = 5.4 \text{ cm}^3 \text{ mil}/100 \text{ in}^2 \text{ 24 hr atm}\}$

Table 8.1 presents activation energies and permeability constants for some polymers and permeants.

Table 8.1   Permeability Constants and Activation Energies [3]

Polymer	Permeant	T (°C)	P x 10^{14} cm^3 cm cm^{-2} s^{-1} Pa^{-1}	E$_a$ kJ mol^{-1}
HDPE				
0.964 density	oxygen	25	3.023	35.1
	carbon dioxide	25	2.70	30.2
	water vapor	25	90	-
LDPE				
0.914 density	oxygen	25	21.6	42.6
	carbon dioxide	25	94.5	38.9
	water vapor	25	675	33.4
Nylon 6	oxygen	30	0.285	43.5
	carbon dioxide	20	0.66	40.6
	water vapor	25	1,328	-
PET				
crystalline	oxygen	25	0.263	32.3
	carbon dioxide	25	1.275	18.4
	water vapor	25	975.	2.9
amorphous	oxygen	25	0.443	37.6
	carbon dioxide	25	2.25	27.6
PVC	oxygen	25	0.340	55.6
	carbon dioxide	25	1.178	56.9
	water vapor	25	2,060	22.9
PVDC	oxygen	30	0.040	66.5
	carbon dioxide	30	0.225	51.4
	water vapor	25	3.75	46.0
PVOH	oxygen	25	0.067	-
	carbon dioxide	25	0.090	-

(cm^3 cm/cm^2 s Pa) x (2.224 x 10^{15}) = cm^3 mil/100 in^2 24 h atm)

8.2.9 Permeability Constants for Multi-Component Structures

8.2.9.1 Planar Multilayer Structures

For multilayer structures, the permeability constant for the material can be estimated from knowledge of the permeability constants of the individual layers, according to the following formula, derived from mass transfer considerations:

$$L_t/\overline{P}_t = \Sigma \, (L_i/\overline{P}_i) \tag{8.21}$$

where t denotes the property of the total structure, and i the individual layers. L, as before, stands for thickness. The formula for water vapor transmission rates is identical, with WVTR substituted for $\overline{P}$. The following examples illustrate application of Eq. (8.21):

Example 6:

A multilayer structure consists of 25 μm HDPE and 50 μm PP. What is the overall permeability constant for oxygen if the oxygen permeability constant for HDPE is 4.0 x 10^4 cm³ μm/m² d atm and for PP is 5.0 x 10^4 cm³ μm/m² d atm?

Answer 6:

$\overline{P}_t$ = 75 μm/[25 μm/(4.0 x 10^4 cm³ μm/m² d atm) + 50 μm/(5.0 x 10^4 cm³ μm/m² d atm)] = 4.6 x 10^4 cm³ μm/m² d atm

Example 7:

A multilayer structure consists of 1 mil polycarbonate, 0.5 mil PVDC, and 2 mil polycarbonate. What is the overall WVTR if the WVTR for PC is 11.0 g mil/100 in² 24 h at 37.8°C, 90% RH, and the WVTR for PVDC is 0.4 g mil/100 in² 24 h at 37.8°C, 90% RH.

Answer 7:

WVTR$_t$ = 3.5 mil/[1 mil/(11.0 g mil/100 in² 24 h) + 0.5 mil/(0.4 g mil/100 in² 24 h) + 2 mil/(11.0 g mil/100 in² 24 h)] at 37.8° C, 90% RH = 2.30 g mil/100 in² 24 h at 37.8°C, 90% RH.

It can be seen that the position of the layers does not affect the result, and in fact, only the total thickness of each material is relevant; the number and thickness of layers the material is divided into is immaterial. This calculation obviously fails to take into account the influence of co-permeants on the material, or the possibility of changes in diffusivity with changes in concentration. Therefore, it fails for situations such as permeation of oxygen through multilayer materials containing EVOH, where the position of the EVOH layer determines how much moisture it will be exposed to, and therefore affects its barrier capability.

8.2.9.2 Cylindrical Multilayer Structures

For cylindrical multilayer structures, such as the sides of a jar or bottle, the relationship of the permeability constant for the overall structure to that of the individual layers is as follows [2]:

$$\ln \frac{R_n/R_o}{\overline{P}_t} = \ln\frac{R_1/R_0}{\overline{P}_1} + \ln\frac{R_2/R_1}{\overline{P}_2} + \dots + \ln\frac{R_n/R_{n-1}}{\overline{P}_n} \tag{8.22}$$

where $R_i - R_{i-1} = L_i$, the thickness of the i'th layer in the composite structure.

8.2.9.3 Blends

Polymer blends can be categorized as miscible if the polymers are mutually soluble, or immiscible if they aggregate into domains of a single polymer type. The morphology of the domains is dependent on processing conditions, relative concentrations, and the nature of the polymers.

The permeability constant of a miscible blend can be calculated as follows [2]:

$$\ln \overline{P}_t = \phi_1 \ln P_1 + \phi_2 \ln P_2 \tag{8.23}$$

where ϕ represents the volume fraction of the component.

Truly miscible blends of polymers are rare. For immiscible blends, the relationships are more complex. For those in which one phase is continuous and the other is dispersed as spherical particles, the following equation may be used [2]:

$$\overline{P}_t = \frac{P_c[P_d + 2P_c - 2\phi_d(P_c - P_d)]}{[P_d + 2P_c + \phi_d(P_c - P_d)]} \tag{8.24}$$

where d represents the discontinuous phase and c represents the continuous phase. This equation does not adequately represent the behavior of other geometries, such as where the discontinuous phase takes the shape of small platelets.

Generally, the simplest method of determining the barrier properties of blended polymers is to test the actual containers, rather than to apply equations based on the properties of the pure materials.

8.2.9.4 Filled Polymers

For polymers with inorganic fillers, provided good adhesion and wettability are present, permeability can be estimated by:

$$\overline{P}_t = \overline{P}_i \phi_i (1 + L\phi_f/2W)^{-1}$$

(8.25)

where i represents the unfilled polymer; ϕ_f is the volume fraction of filler; and W/L is the aspect ratio of the filler [2]. The aspect ratio is defined as the average dimension of the filler particles parallel to the plane of the material divided by the average dimension perpendicular to the film. It is assumed there is no migration through the inorganic phase. If there is good adhesion between the polymer and the filler, permeability decreases with increasing filler concentration. However, if adhesion is poor, interphase microvoids can form, which increase permeation, and is not accounted for in Eq. (8.25).

8.2.10 Permeation of Organics

Permeation of organic compounds is a concern when plastics are used to package these compounds, either alone or as a solvent for another product. In addition, organic compounds are often associated with the odor and flavor of a product. A product can lose desirable odors or flavors to the surroundings, or can gain undesirable odors and flavors from the surroundings. Even small losses or gains of aroma compounds are a concern, and measurements of permeability coefficients can be very difficult under these conditions, especially for plastics which are relatively good barrier materials. However, commercial equipment for measuring transport of organic materials is now available [6].

Unlike permeation of oxygen, nitrogen, and other gases, measurements of permeation of organics, especially odors and flavors, must often be done at non-steady state conditions, and results may require lengthy test periods. For many odor and flavor compounds, the shelf life of the product is shorter than the time required to reach steady state transfer of the compound. Elevated temperatures are frequently employed to speed up measurements, but for polymers which are normally below their glass transition temperature, care must be taken to remain below that temperature during these measurements.

An additional concern with predicting migration of organics is that permeability constants are often not truly constant, but are affected by the concentration of the organic within the polymer. Interaction between the organic permeant and the polymer can cause swelling and other changes in polymer morphology which significantly affect diffusion rates, and thus permeability. For some polymers, permeability constants are also affected by relative humidity, for similar reasons. In some cases, therefore, co-permeation of the organic and water vapor, or of multiple organics, needs to be considered in order to arrive at accurate measures of permeation and calculations of shelf life.

Compared to the information available about oxygen, carbon dioxide, and water vapor permeation, relatively little data is available about the permeation of many organic compounds significant to the packaging industry. DeLassus has tabulated permeability constants, diffusion coefficients, and solubility coefficients for several flavor and aroma compounds in vinylidene chloride copolymer, EVOH, LDPE, HDPE, and PP [1]. All measurements are reported at 25 °C and "dry" conditions, and some are extrapolated from measurements performed at higher temperatures.

8.2.11 Concentration Dependence of Permeability Constants

As was mentioned earlier, in some systems the permeability constant is a function of the concentration of the permeant, or of some other substance such as water vapor, in the polymer. Generally the effect, if any, is to increase the diffusion rate for the organic, oxygen, or other gases, as the concentration of organic in the plastic rises, since the organic molecule can plasticize the polymer. Similarly for water-sensitive polymers, water can act as a plasticizer, increasing the diffusion coefficient of oxygen and other gases, as well as of organics.

In such instances, unless a steady state is reached, it is generally necessary to use numerical methods or direct measurement to determine the barrier capability of the polymer, since the mass transport equations cannot be solved analytically. Indeed, often little is known about the value of the diffusion coefficients as a function of concentration, making it impossible to do accurate calculations, even using numerical methods. This is particularly true when the permeation involves multiple organic compounds, as is often the case for odors and flavors.

Fortunately, the concentration dependence of diffusivity and hence of permeability is, in general, less of a concern for odor and flavor components than for organics present in higher concentrations. The very small concentrations of aroma compounds typically present in the polymer usually do not produce significant changes in diffusivity.

Concentration-dependent diffusion is known as non-Fickian diffusion. Some theoretical discussion and several examples of measurement of co-permeation of organics and of organics and water vapor has been presented by Giacin and Hernandez [7].

8.3 Migration

Migration refers to the transfer of materials from a package to the package contents. Migrants can include residual monomer, catalysts, solvents, antioxidants, and other polymer additives. Migrants may be of concern for health reasons, such as residual vinyl chloride in PVC; or may be related to off-flavors or odors in the product. Sometimes migration is desirable, as when a migrating antioxidant from a plastic film provides protection against oxidation in the product. More often, migration is not desirable. Even extremely tiny amounts of some compounds can produce noticeable flavor or odor changes in certain products. Migration is very similar to permeation in that Fick's laws of diffusion are generally governing. In this case however, the amount of migrant is fixed, so the concentration in the polymer phase decreases with time. Also, depending on the situation, migration may be from the container into the product only, or may be both to the product and to the external surroundings. Also, depending on the product, diffusion within the product may or may not play a significant role. The partition coefficient for the component of interest describes the relative amounts of the migrant in the polymer and the product when equilibrium is reached. It can be defined as [2]:

$$K = c_f/c_p \qquad (8.26)$$

where c_f is the equilibrium concentration in the product and c_p is the equilibrium concentration in the polymer. Hernandez gives a number of examples of solutions to Fick's law in representative cases of migration [2].

At equilibrium, the concentration of migrant in the product is a function only of the initial concentration in the polymer, the partition coefficient, and the ratio of polymer and product [2]:

$$C_f = \frac{C_p^o}{1/K + V_f/V_p} \qquad (8.27)$$

where C_f is the equilibrium concentration in the product; C_p^o is the initial concentration in the polymer; V_f is the volume of product; and V_p is the volume of polymer.

Recently, there has been a great deal of interest in potential migration of contaminants from recycled plastics used in food contact packaging, and particularly in the ability of a functional barrier layer in the package structure to prevent such migration. Various models for migration in such situations have been proposed. The F.D.A. currently accepts the model proposed by Begley and Hollifield [4] for the purpose of determining whether a proposed use of recycled plastic merits a "letter of nonobjection" from FDA. This model assumes Fickian diffusion, a constant concentration of contaminant in the polymer, and no resistance to the transfer of the contaminant from the polymer to the product. Thus, it significantly overestimates potential migration. Laoubi and Vergnaud recently presented a model which takes into account the coefficient of mass transfer at the packaging-food interface [5]. Both models employ dimensionless numbers, simplifying the use of their results in a variety of packaging situations.

8.4   Scalping

Scalping refers to migration of a component from the product to the package. These are frequently flavor, aroma, or color compounds, and the loss is generally detrimental to the product. As with migration, Fick's Laws can be used to analyze the mass transfer, and the partition coefficient indicates the equilibrium concentrations in the product and package of the components of interest.

In this case also, the equilibrium concentrations of the component depend only on the initial concentration (in this case the concentration in the product), the partition coefficient, and the relative volumes of the product and package [2]:

$$C_f = \frac{C_f^o}{V_p/KV_f + 1} \tag{8.28}$$

where C_f is the concentration in the product at equilibrium; C_f^o is the initial concentration in the product; and V represents volume as described above. However, the product may reach the end of its shelf life in terms of unacceptable loss long before equilibrium is reached.

The situation becomes more complicated when the compound is able not only to dissolve in the polymer, but also to migrate through it and be lost to the outside atmosphere. The analysis must then include solubility and diffusivity, along with the partition coefficient and the vapor pressure of the component.

8.5   Modified and Controlled Atmosphere Packaging

In recent years, there has been significant growth in the use of modified and controlled atmosphere packaging to provide extended shelf life for products. In many cases, this involves the use of plastic packaging materials. The idea is to provide to the product an atmosphere that differs from ordinary air, so that reactions, usually with oxygen, can be slowed down and shelf life improved. The basic difference between modified and controlled atmosphere packaging is that in controlled atmosphere packaging, there is active continuous control over the atmosphere, to maintain it at some desired composition. Modified atmosphere packaging, in contrast, begins with something other than normal air in the package, but the atmosphere in the package changes with time, influenced by the reactions of the product and by permeation of the package material. Controlled atmospheres are associated more with storage facilities, or sometimes large bulk packages, than with individual retail-type packages, while individual packages are most likely to use modified atmospheres.

For many oxygen-sensitive products, in particular fresh produce, meat, and poultry, shelf life can be significantly extended by limiting the exposure of the product to oxygen. For fresh meat, this can be accomplished by vacuum packaging in a high barrier material. However, the oxygen-deprived fresh meat takes on a dark purple color; in the U.S., consumers have not been confident of these products, since they associate a bright red color with high quality. Nonetheless, such packages are used widely for the distribution of large cuts of meat, and are beginning to be found in retail packages as well.

For fresh produce, this solution is not feasible. Fresh produce in a vacuum package rapidly decays, as the normal aerobic metabolism of the tissue is replaced by fermentative metabolism when oxygen content is lower than 2 to 4% [8]. Thus, to maintain quality, the product must respire to keep the cells living. However, respiration also brings aging. Therefore, the longest shelf life is obtained when respiration is slowed but not stopped.

One way to decrease respiration, for some products, is with an atmosphere low in oxygen and high in carbon dioxide. Carbon dioxide has the further effect of suppressing the growth of many spoilage microorganisms. In addition to direct antimicrobial effects, CO_2 is readily soluble in water, fats, and oils, including those in produce. When it dissolves, it produces an acid solution, further retarding microbial growth [8].

Nitrogen is often used as a component of modified atmospheres because of its abundance and hence low cost (air is 78% nitrogen), and because it is essentially inert in most packaging conditions. Other compounds such as carbon monoxide, sulfur dioxide, ethanol, and argon have seen limited use in such packages [8]. The precise mixture of gases for optimum shelf life is highly dependent on the particular product.

Once the desired atmosphere is obtained and the package is sealed, it often does not stay constant. For example, the product, if respiring, continuously converts oxygen into carbon dioxide. Therefore, what is desired is a package that transmits both oxygen and carbon dioxide at the rates at which they are respectively consumed and created, to maintain a constant atmosphere in the package. Even in the case of products such as baked goods where the desire is to maintain a high concentration of carbon dioxide and prevent entry of oxygen and loss of water vapor, the atmosphere inside the package changes with time unless the package is a perfect barrier. Nonetheless, very significant increases in shelf life can be obtained.

Modified atmosphere packaging is particularly useful for fresh prepared foods, which generally have a very short shelf life. This technique has made it possible for consumers to purchase products such as prepared tossed salads, ready-to-eat carrots, and fresh pasta at economical prices. For fresh produce, such an atmosphere can be provided during packaging, or can be allowed to develop by itself as the product respires within a package that provides a high enough barrier that oxygen is consumed and carbon dioxide generated by respiration faster than they permeate through the plastic material. Selection of an appropriate polymer to provide the desired balance of oxygen and carbon dioxide permeability can be very complex. The goal is typically an atmosphere containing 2 to 10% oxygen and carbon dioxide. For products such as mushrooms and peas, which have a high respiration rate, even low barrier films such as LDPE and PVC are not permeable enough to avoid depletion of oxygen. Microperforated films or larger pores coupled with a breathable label have been utilized. Some experimentation has also gone on with oxygen mixtures as high as 70 to 100%, which have also been found to inhibit spoilage in some produce [8,9].

The control of water vapor inside packages has long utilized desiccant technology, in which a small packet of desiccant is placed inside the package to absorb water vapor which permeates through the package. Similar sachets are now available for the removal of oxygen. They generally contain iron, which in the presence of oxygen and water vapor oxidizes (rusts), consuming the oxygen and removing it from contact with the product. These oxygen absorbers are now available incorporated into labels and other package components, so that a separate sachet is not necessarily required. This is particularly useful where there are concerns about accidental ingestion of the sachet. Currently carbon dioxide absorbers are used in controlled storage rooms, but have not yet been adapted for modified atmosphere packaging. Potassium permanganate and activated charcoal have been used in sachets and other forms for the purpose of absorbing ethylene and other gases [8].

A new development is the production of films reported to have ethylene absorbers built into the film structure, typically in the form of a ceramic material embedded in the film, although the efficacy of such materials has been seriously questioned. Films with embedded oxidizable compounds for reducing oxygen gain have also been designed, and work on incorporating antimicrobial agents directly into films is also going on [8].

References

1. DeLassus, P. (1997) In *The Wiley Encyclopedia of Packaging Technology, 2nd ed.*, A.L. Brody and K.S. Marsh (Eds.), New York:Wiley, pp. 71-77
2. Hernandez, R.J. (1996) In *Handbook of Plastics, Elastomers, and Composites*, 3rd ed., C.A. Harper (Ed.), New York:McGraw-Hill, pp. 8.1-8.63
3. Brandrup, J. and Immergut, E.H. (1975) *Polymer Handbook*, New York:Wiley
4. Begley, T.H. and Hollifield, H.C. (1993) Food Technology, pp. 109-112.
5. Laoubi, S. and Vergnaud, J.M. (1995) Packaging Technology & Science, *8*, pp. 97-110.
6. Seeley, D. (1997) In *The Wiley Encyclopedia of Packaging Technology, 2nd ed.*, A.L. Brody and K.S. Marsh (Eds.), New York:Wiley, pp. 38-41
7. Giacin, J.R. and Hernandez, R.J. (1997) In *The Wiley Encyclopedia of Packaging Technology, 2nd ed.*, A.L. Brody and K.S. Marsh (Eds.), New York:Wiley, pp. 724-733
8. Day, B.P.F. (1997) In *The Wiley Encyclopedia of Packaging Technology, 2nd ed.*, A.L. Brody and K.S. Marsh (Eds.), New York:Wiley, pp. 656-659

9 Environmental Considerations

During the last 15 years or so, plastic packaging materials have been under attack on environmental grounds. Alhough public concern and legislative pressure has diminished somewhat in the U.S. more recently, packaging's impact on the environment continues to be an issue. The European Union is currently establishing producer responsibility requirements for packaging manufacturers, modeled to a large extent on the system in place in Germany. Canada has a policy to reduce packaging waste, with regulations to ensure goals are met. In much of Asia, packaging as an environmental issue is just becoming an issue. Thus, all around the world, packaging is questioned on environmental grounds, based on a number of perspectives. In this chapter, we examine some of the concerns, their results, and the developing philosophy of lifecycle assessment as a way to reach sound environmental decisions. Much of the detail and examples focus on the situation in the U.S., although references are made to other parts of the world, as well.

9.1 Energy

Beginning in the mid 1970s, the use of plastics in packaging came under increased scrutiny from an environmental perspective. First were concerns about energy use, and specifically about the use of petrochemicals for plastics production. When the Arab oil embargo of 1973 caused people to consider the finite nature of petroleum resources, there was a perception that the use of these valuable energy resources for plastics was perhaps not justifiable. There were calls, for example, for returns to glass in markets where plastics had made significant inroads, such as in beverage packaging.

However, two facts soon emerged. First, less than 2% of U.S. petroleum consumption is used for all plastics production, not just packaging [1]. Second, for many applications, the overall energy consumption for the manufacture and use of plastics packaging is significantly less than that needed to produce and use alternative materials. The easy processability of plastics and their light weight result in energy savings in forming and transportation. Alternatives such as glass and aluminum involve energy-intensive manufacturing steps. Furthermore, plastics can be incinerated and the energy value recovered if they are burned in facilities designed to do this. In any case, environmentally-related concerns about the petrochemical base of plastics decreased as new sources of oil were discovered and prices fell.

9.2   Solid Waste

In the mid 1980s, concern about municipal solid waste disposal became the most pressing environmental issue facing the packaging industry. A number of U.S. cities, particularly on the east coast, experienced significant problems with lack of landfill capacity. Sanitary landfills had made a significant contribution to health and safety in the 1960s and 1970s by replacing open dumping of garbage and periodic uncontrolled burning. In a sanitary landfill, garbage is covered with a layer of soil at the end of every working day. However, many municipal landfills were found to be significant sources of groundwater pollution, as well as air pollution. Many existing landfills were closed, either because of pollution or because they had reached capacity, but new landfills were not being opened as rapidly as needed. Despite regulatory changes to reduce potential pollution, landfills were perceived by the public as highly undesirable neighbors, which made it difficult to open new landfill sites. Incineration with energy recovery was proposed as a major alternative to reduce reliance on landfills, but incinerators proved to be even less politically popular than landfills, with the primary concerns air pollution and ash disposal. When legislators at various levels of government began looking at approaches to reduce waste and so lower the need for waste disposal systems, many quickly focused on packaging.

Studies of municipal solid waste reveal that packaging accounts for approximately one third of the waste stream by weight (see Fig. 9.1). The proportion of plastics of all types has increased over the years, but still accounts for less than 10% of municipal solid waste generated by weight in the U.S. (see Fig. 9.2). Packaging plastics make up about 48% of the plastics fraction [2]. These numbers have been used to argue that plastics are a relatively insignificant component of the waste stream.

However, when looked at by volume, the relevant variable in terms of landfill capacity, a somewhat different picture emerges. While precise figures are not available, the U.S. Environmental Protection Agency estimated plastics comprise approximately 24% of the total volume of municipal solid waste sent to disposal (materials recycled or composted are subtracted from materials generated before this calculation), with plastic packaging accounting for just over 11% of the total volume (Fig. 9.3) [3]. Thus, plastic packaging appears to be much more significant in terms of volume than weight. However, it has also been shown that the amount of packaging sent to disposal would increase substantially if alternatives to plastic, such as glass and metal, were used.

In response, a number of legislators called for bans on various types of plastics packaging, or requirements that degradable plastics be used in place of nondegradable ones. However, it was soon shown that in properly designed landfills, the dry anaerobic environment retarded

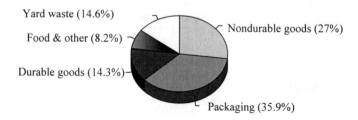

Figure 9.1   Product categories of municipal solid waste in the U.S. in 1994, by weight [2]

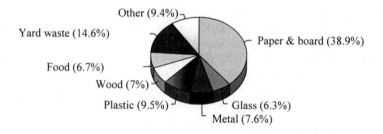

Figure 9.2   Material categories of municipal solid waste in the U.S. in 1994, by weight [2]

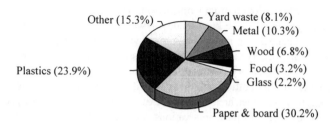

Figure 9.3   Material categories of municipal solid waste in the U.S. in 1994 after recycling, by volume [3]

microbial growth, and consequently, biodegradable materials took a long time to break down. In sum, using biodegradable packaging materials would do almost nothing to extend the useful life of landfills. In fact, in many cases, using biodegradable materials would be counterproductive, since the replacement materials were bulkier than the plastic alternatives and would fill the landfills up more rapidly. There are, however, some applications where degradable plastics make sense; these are discussed later.

With this increased understanding of the limits of degradability came increased attention to the remaining alternatives for reducing the presence of plastics (and other packaging materials) in the waste stream - source reduction, reuse, and recycling.

9.3 Source Reduction

Source reduction refers to the use of less packaging material to accomplish the same task of delivering product to the user. If less packaging is used, less enters the waste stream. Source reduction has been practiced by the packaging industry for years for economic reasons. Source reduction often involves producing lighter weight containers (see Fig. 9.4) and switching from rigid containers to flexible packaging. These changes often include changing packaging materials from glass, metal, or wood to plastic.

Generally, if less packaging material is used, money is saved. Of course, if, as a result of decreased packaging, product damage increases, then there may be no savings at all. Thus, a balance must be forged between the use of more packaging to provide more protection, and the use of less packaging to save money. As a general rule, expensive products are packaged to afford a higher degree of protection and less damage than less expensive products. Similarly, products such as hazardous materials, where the consequences of their release to the environment can be severe, are packaged for a higher degree of protection than innocuous products. Source reduction, in practice, is also limited by the requirements of the distribution system, marketing considerations, legal requirements, and other factors.

Source reduction explicitly for environmental reasons has increased in recent years. Some of the packaging changes might have been made anyway for economic reasons, but some occurred as a result of changes in public perception of value. For example, many products, especially in the health and beauty aids category, no longer have paperboard cartons holding bottled products and tubes. Laundry detergent is now sold in thin-walled, no-frills plastic bottles as refills for the sturdier bottles with measuring devices built into the caps. Products have changed as well; there has been an increased use of concentrates that can be packaged in smaller containers. The success of these innovations depends on the public's willingness to buy something less than they used to buy, because they see these alternatives as environmentally positive.

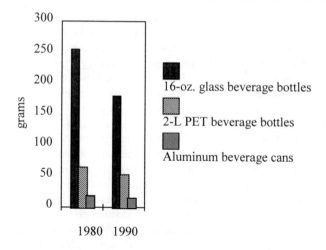

Figure 9.4   Changes in the average weights of containers

In 1991 the Packaging Task Force of the Coalition of Northeastern Governors (CONEG), mentioned in Chapter 7, published the *Preferred Packaging Manual* which promulgated a set of guidelines for reduction of packaging waste, specifying source reduction as the best alternative, although reuse and recycling were included. Their guidelines for package design were as follows:

1. Eliminate packaging whenever possible
2. Minimize packaging
3. Use packaging which is consumable, returnable, or refillable/reusable
4. Use recyclable packaging and recycled content in packaging [4]

At the same time CONEG issued the "Packaging Challenge" in which the top 200 users and producers of packaging in the U.S. were asked to voluntarily commit themselves to implementing the packaging guidelines, set goals for reducing packaging toxicity and waste, measure progress in achieving these goals, and report their progress to CONEG [4]. Several companies accepted this challenge, although most did not.

In 1992, CONEG formulated "Model Source Reduction Legislation" which was intended, as was the Model Toxics Legislation, to be easily adoptable by various states and facilitate legislating packaging reduction. At the time CONEG felt that by 1996 a 15% reduction in packages or packaging components entering the solid waste stream from 1988 levels was feasible, and suggested a study to see if a 35% reduction by 2000 was possible. Companies could calculate their reduction on either packages sold or distributed in individual states, or in the U.S. as a whole. They could use either a company-wide approach or a specific package approach. If the company-wide approach was adopted, total elimination of packages or package components was counted at 1.5 times the actual reduction, as was source reduction which enhanced or maintained the recycling rate for a package, under certain circumstances. Use of recycled content counted as much as actual reduction in weight.

In the specific package approach, packages or components needed to be either reduced in weight by 10%, be reusable or refillable, contain at least 25% recycled post-consumer materials, or have a recycling rate of at least 25%[5]. As of 1996, no states have adopted the Model Source Reduction Legislation, although some states have passed similar legislation (see Section 9.9.3). Further, the CONEG Packaging Task Force was disbanded in 1996.

Despite the successes in source reduction, it is a largely invisible alternative. People cannot usually tell a package weighs less. Changes in package type which result in source reduction soon become standard. It is difficult to document, as it is hard to count, the amount of material that you do not use. Further, source reduction provides limited opportunity for individual consumers to feel like they are doing something about the "solid waste problem." For these and other reasons, recycling has attracted considerably more public attention and enthusiasm than source reduction.

9.4 Reuse

One way to decrease the amount of packaging entering the waste stream is to use reusable packaging, which is intended to provide more than one product containment cycle without remanufacturing. In general, reusable packaging must be stronger than singe-use packaging, and therefore more material is needed per package. However, the overall savings can be considerable if the package has a significant number of use cycles before disposal. Savings in energy and raw materials can also often be achieved.

Unfortunately, the use of refillable packaging for consumer items is quite limited. Some refillable PET soft drink bottles are used, though not in the U.S. There is very limited use of refillable bottles for shampoo and similar products, mostly for products sold by beauty salons, and by one store chain in Germany. The use of refillable plastic bottles for retail store-dispensed drinking water is more common, but still rare. Refillable polycarbonate milk bottles have been used successfully in a few school lunch programs. On the whole, however, it has proved difficult to institute reliable, efficient, and economical return systems for the packages. Contamination of the containers, and thus potentially of the product, is also a concern.

In contrast, the use of refillable distribution packaging has grown considerably. The soft drink, dairy, and bakery industry have made considerable use of reusable packaging, often plastic, for a number of years, purely for economic reasons. More recently, the major automobile manufacturers have found use of refillable, often plastic, packaging for parts distribution to be of significant value in reducing their costs for solid waste disposal, and also in maintaining a clean environment inside the plants. Although plastic crates or bins are initially more expensive than the corrugated fiberboard boxes they often replace, the overall costs, including the avoided waste disposal costs, are generally less, providing a potent economic incentive for making the changes. The furniture industry, especially manufacturers of office furniture, has made similar changes for similar reasons. Plastic pallets are increasingly replacing wood ones; they have a higher initial cost but a much longer service life. Other examples could be cited, as well.

Key to the use of refillable packages is a reliable mechanism of returning packages. A method of tracking the returned packages, especially if they leave the control of the business entity which owns them, is often necessary, since the packages represent a significant investment.

A number of mechanisms have been devised for returning the empty packages. For auto parts and furniture distribution, for example, the trucks delivering the filled packages can backhaul the empties, either from the same or from a previous shipment. Users and manufacturers of expanded PS packaging have sometimes enclosed shipping labels providing for the return of the empty package by UPS or another carrier, free to the consumer, for purposes of either recycling or reuse. Loosefill plastic packaging is collected for reuse by a number of businesses that provide shipping services.

9.5   Recycling

Recycling enjoys a great deal of political popularity, despite recent criticism, some of which is based on economics and environmental effects, such as energy use. Recycling has become part of ordinary life for many people, and they generally do not want to give it up.

Recycling involves taking materials which would otherwise enter the waste stream and reprocessing them into useful products. Recycling can be classified in a number of ways. One such division distinguishes process scrap, industrial waste, and post-consumer waste. Process scrap is material generated during plastics production, such as flash from the extrusion blow molding of bottles. These materials are routinely reground and fed back into the process, and their use is not generally considered recycling. Industrial waste covers a variety of waste materials generated during the process of fabricating packages, which were generally discarded before recycling opportunities were developed. An example is "off-spec" (defective in some way) in-mold paper-labeled bottles. These bottles could not simply be ground up and fed back into the process, since the paper was an unacceptable contaminant. Until processes were developed which removed the labels and cleaned the material, these bottles were destined for disposal, either by landfilling or perhaps by incineration with recovery of the energy content of the materials. Post-consumer waste refers to packages which have completed the normal product distribution cycle and reached the end user. The end user may be an individual consumer or a business entity. The key part of the definition is that the package has fulfilled its intended purpose and is now a waste material.

Another classification of recycling focuses on the use of the material, rather than its origin. In this classification scheme, primary recycling refers to the use of the recycled material in an application identical or similar to its original one. An example is the recycling of HDPE laundry detergent bottles, into other laundry detergent bottles or into motor oil bottles. Secondary recycling refers to recycling into an application which has less stringent specifications than the original, implying downgrading the material. An example is recycling HDPE bottles into traffic cones. Tertiary recycling refers to the use of the recycled material as a chemical feedstock. For example, PET can be converted to monomers by glycolysis and then repolymerized to PET. Finally, some people classify use of the recovered material as an energy source as quaternary recycling, while others argue that this should not be considered recycling at all.

Still another classification of recycling is "closed loop" or "open loop." Closed loop recycling refers to the use of the recycled material in its original application, regardless of the process used to get it there. Thus, the tertiary recycling of PET beverage bottles is closed loop recycling if the repolymerized PET were used again in beverage bottles. Open loop recycling refers to all types of recycling where the reuse is different from the original use. For example, the use of HDPE milk bottles to make HDPE laundry detergent bottles, the detergent bottles to make recycling bins, and the bins to make landscape timbers are examples of open loop recycling.

Many people have strong opinions as to which types of recycling are better than others. The most vocal seem to be those who hold that closed loop recycling is the best, and those who hold that recycling materials into long life-time applications (which generally means open loop recycling) is preferrable. A more balanced view is that the best method provides the greatest reduction in the use of virgin material, with, of course, other environmental impacts considered as well. It is hard to understand why the use of 100 tons of recycled PET to make beverage bottles and 100 tons of virgin PET to make carpet is better or worse than the use of 100 tons of recycled PET to make carpet and 100 tons of virgin PET to make beverage bottles, if all else in the system is equal. Of course, in this particular example, the energy requirements associated with use of the recycled PET in beverage bottles are generally greater than those associated with its use in carpet, because of the increased purity required.

9.5.1 Collection of Materials

Collection of plastic packaging materials for recycling takes three basic forms. The most successful in terms of percentage recovery of materials is deposits. Ten U.S. states (Table 9.1) have some form of deposit legislation on beverage containers. In most cases, the consumers pay a deposit of $0.05 per container, charged separately from the beverage cost. Covered

Table 9.1 Beverage Container Deposit States

California	Connecticut
Delaware	Iowa
Maine	Massachusetts
Michigan	New York
Oregon	Vermont

beverages are carbonated soft drinks, beer, and sometimes wine coolers. The customer can take the empty containers back to the retailer for a refund of the deposit. Deposits on large containers are sometimes greater. In Michigan, deposits are $0.10 on nearly all containers. In Maine, beverages such as juices are also included in the deposit. California has a system which is not a true deposit, in that no separate charge is made to the consumer for the refund value of the container. The consumer can get a refund only at designated redemption centers, rather than at any retailer. Current refund values in California are $0.025 per container. The plastic containers used for covered products are almost exclusively PET. Return rates in general exceed 90%, although they are significantly lower in California, as a result of the combination of lower economic incentive and decreased convenience in returning the containers.

Beverage manufacturers and retailers are generally strongly opposed to bottle deposit legislation (which they term "forced deposits") and have vigorously, and with considerable expense, campaigned against them. They have been successful in defeating attempts to expand the deposit concept to additional areas, with the exception of California's compromise system. Their argument is that the costs of deposit systems are excessive, and there are better, more economical ways to accomplish the goals, particularly that of recycling. Nonetheless, surveys in states with deposit legislation show residents to be generally satisfied, and no deposit legislation has been repealed. On the contrary, deposits have been expanded to include new categories of containers in several states. In evaluating the effectiveness of deposit systems, it must be kept in mind that the major purpose of these laws is not to promote recycling, but to combat litter. A number of studies have shown bottle deposit legislation to be effective in reducing the number of beverage containers discarded along highways, in parks, and other locations, although it is not effective in reducing other types of litter.

Whether deposits are the most efficient economic means to accomplish this goal is more debatable. It is also true that a deposit system can quickly become unwieldy if it is extended to a large variety of plastic containers in efforts to promote widespread collection for recycling.

Another alternative for collection of materials is drop-off sites. These depend on the voluntary cooperation of individuals to bring materials to some centralized collection point. This is usually the least expensive type of recycling collection to set up and operate. The basic requirement is a container of some type, arrangements for periodic collection of the material by the recycler from the collection point, and some minimal publicity to let the prospective users know about its availability. Publicity may be as simple as a sign above the container explaining what should be placed inside. There may be only one collection site for a whole municipality, or there may be a number of collection sites. The percentage of available material diverted from the waste stream by this type of collection varies considerably according to the convenience of the sites and the amount of effort expended on educating people and convincing them to participate. In general, participation rates are quite low, on the order of 10 to 20%, or even less. Participation rates are measured by the fraction of consumers who place some materials in the drop-off collection sites, and are greater than diversion rates, since few consumers recycle 100% of the recyclables they generate.

For plastic packaging, two different types of drop-off collection systems often co-exist. In the U.S., centralized sites are most common for plastic containers, and PET and HDPE bottles are usually the only types accepted. A number of retail chains, especially grocery stores, maintain collection barrels for PE shopping bags, and less frequently for expanded PS containers. Some drop-off operations increase participation by offering a monetary payment to consumers for the material they bring. This has been a staple of the aluminum can recycling system in non-deposit states, but is seldom very applicable to plastics because of their low value per container. In fact, with the growth of curbside recycling programs, these buy-back programs have largely been discontinued, with the exception of those for aluminum cans.

The third alternative is to collect materials from consumers, rather than require them to bring the materials to a collection point. This offers a significant advantage in increasing convenience to consumers, and consequently increasing participation. In the U.S., the most common form is curbside collection, where individual households place their recyclables at the curb, usually at the same time as garbage collection. The most effective systems involve providing a bin for the household to use for the recyclables, and allowing them to commingle various types of recyclables in the bin. The recyclables are sorted either at truckside or in a processing center. Distributing a bin, often blue in color, has proven to be a significant factor in the degree of success of a program. It not only gives the householder a convenient place in which to put the recyclables and serves as a visible reminder of the recycling program, but also mobilizes peer pressure to encourage recycling. It is easy to see by looking down the street on garbage day who has their recycling bin out and is thus a "good citizen" and who does not. Good curbside collection programs, with associated publicity and education, can have participation rates of 80% or more. Most programs that achieve rates this high are in communities that mandate participation by residents, but some have reached these levels with voluntary programs. Education programs for local school children are particularly influential in getting recycling habits started early.

For multi-family residences, recycling programs are less commonly available than for single family homes. Communities often maintain a system of drop-off recycling programs to serve these residents, in tandem with curbside programs. In the best of such programs, these drop-off sites are right in the complex with the multi-family homes, often in a laundry room or other central area.

Collection programs targeted at commercial users rather than individual households are very significant for recycling LDPE film, particularly stretch wrap. Factories, warehouses, and other facilities which generate large quantities of PE are the major sources of recycled LDPE.

In most programs for plastics recycling, plastic packaging materials are not the only materials collected. For example, curbside programs typically collect newspaper, glass containers, aluminum packaging, and sometimes other materials, as well.

9.5.2 Status of Plastics Recycling

PET is currently the plastic packaging material with the highest recycling rate. In the U.S. 48.6% of PET beverage bottles were recycled in 1994 [6], and for all PET bottles the recycling rate was 34%. There is little, if any, recycling of non-bottle PET packaging, but bottles make up about 92% of rigid PET packaging. Excess capacity for producing virgin PET and consequent low prices for both virgin and recycled grades severely impacted the economics of PET recycling in 1996, and continued into 1997.

HDPE is the plastic next most often recycled, with a rate of nearly 26% for unpigmented bottles and 10.8% for pigmented bottles in 1994 [7]. The overall recycling rate for HDPE containers was estimated at 11% that year [8].

Other plastic packaging materials are recycled to a lesser extent. LDPE merchandise bags were recycled at a rate of 10 to 15% in the U.S. in 1994, and the overall recycling rate for plastic film was about 2% [9]. In the U.K., on the other hand, PE film is reported to be the most widely recycled plastic packaging material [10].

Polystyrene recycling was established in the mid-1980s, largely in response to the public's perception that PS was a significant litter and solid waste problem. In the U.S., the National Polystyrene Recycling Company, which focuses on food service polystyrene, and the Association of Foam Packaging Recyclers, which focuses on cushioning material are primarily involved in PS recycling. A recycling rate of 10% was reported for 1995 [11].

PVC recycling occurs more in Europe than in the U.S., because of its wider use there. The U.S. recycling rate for PVC bottles in 1992 was reported to be only about 1% [12]. In France, recycling rates for PVC bottles are betweeen 20 and 47% [13]. Polypropylene and most other packaging plastics are recycled primarily in programs that collect mixed plastic bottles. Recycling rates are very low for these materials.

9.5.3 Uses for Recycled Plastic Packaging

Recycled plastics find new lives in a variety of forms, many in non-packaging applications. The earliest large market for recycled PET was polyester fiberfill for sleeping bags, jackets, and similar uses. Newer fiber markets include carpet and clothing. Non-fiber markets include automobile distributor caps, strapping, produce trays, and even soft drink bottles. In fact, three different approaches to recycling PET into food packaging have been accepted by the FDA.

One method is sandwiching a layer of recycled PET between layers of virgin resin, which provide a barrier to the migration of contaminants. A second approach is a tertiary recycling process in which the recovered PET is broken down to monomers by glycolysis or methanolysis, purified by crystallization, and then repolymerized. This material can then be used in direct food contact applications. For economic reasons, this repolymerized PET is generally blended with virgin resin in proportions of 25% recycled to 75% virgin. The third method involves control over source materials (generally deposit bottles) and intensive physical processing to ensure removal of most contaminants. This material can also be used in direct food contact applications.

The first large volume market for recycled HDPE was in agricultural drainage pipe, where it competed with off-spec resin, which does not meet the manufacturers' usual standards. However, the use of recycled HDPE in bottles has now surpassed its use in pipe. Most liquid laundry detergent bottles in the U.S. contain a sandwiched layer of recycled HDPE and regrind between layers of virgin HDPE. Motor oil bottles are usually a blend of recycled HDPE and virgin resin. Recently a multilayer HDPE bottle containing an enclosed layer of recycled resin was approved for limited food contact applications [14]. Another significant market for recycled HDPE is in film production, particularly for merchandise bags. Pallets and plastic lumber also consume significant quantities of recycled resin. Nonpackaging uses include traffic cones and barriers, flower pots, toys, recycling bins, and other household products. HDPE base cups from beverage bottles are most often used in the manufacture of new base cups, again blended with virgin resin, although this market is shrinking because PET bottles are increasingly being made without base cups. A major market for recycled LDPE is trash bags, followed by merchandise bags, agricultural film, envelopes, housewares, and plastic lumber.

Recycled PS is used for housewares, produce trays, egg cartons, and cushioning materials. Some recycled PS is used in foam insulation board for construction applications. Sandwiched layers of PS have been used in fast food containers, and undensified PS foam is used as fill for lawn furniture, stuffing for flower vases, soil lightener for plant nurseries, and as a drainage medium for ground water.

Recycled PVC also finds a variety of applications, including pipe, household items, pickup truck bedliners, and packaging. Recently recycled PVC has also been used in clothing and in accessories such as fanny packs.

Markets for mixed recycled plastics are more limited than for single resin materials, since properties suffer as a result of the mutual insolubility of most plastics. Therefore, most markets are for items with thick cross-sections and relatively undemanding applications. Many of these fall in the general realm of wood and concrete substitutes. Plastic lumber has a significant advantage over wood in outdoor applications, where wood tends to eventually rot and plastic has a much longer useful life. Plastic "wood" may cost three times the price of wood alternatives, which has limited its widespread use, even though the total life cycle costs are often lower. However, as environmental concerns about treated wood increase, plastic alternatives are likely to become more attractive. Applications include fencing, boat docks, and landscape timbers. Plastic railroad ties are being tested. Plastic can substitute for concrete in flooring, machinery bases, and parking stops. Investigators at Rutgers University, Piscataway, New Jersey, used commingled plastics plus compatibilizers and other additives to produce a resin reported to be suitable for blow molding containers [15]. For most high value applications, however, sorting the recovered plastics by resin type is key.

9.5.4 Sorting Plastics by Resin Type

The simplest approach to production of single-resin, recycled plastic streams is to collect only the desired materials, so that the stream is specific and well characterized. The first recycling programs for packaging plastics focused on PET beverage bottles or HDPE milk bottles. The first programs collecting milk bottles instructed consumers to remove the caps and rinse the bottles, and in some cases even remove the labels from the bottles. If instructions were followed, the result was a pure, unpigmented HDPE stream. Of course the problem was that the instructions were never completely followed, particularly when consumers were asked to remove labels. It soon became evident that these expectations were unrealistic if high diversion rates were desired.

PET bottles initially were collected in states with bottle deposit legislation. Processing was more complicated, since these bottles usually had HDPE base cups and aluminum caps. Some systems removed the base cups and closures before the PET bottle was ground and washed. More commonly however, the bottles were ground, air-classified to separate the lighter materials, and washed. Then, the HDPE was separated by density in a float-sink tank or a hydrocyclone and the aluminum separated in an electrostatic separator.

Since then, changes in bottle design have decreased the fraction of bottles with base cups, and aluminum caps have largely disappeared in favor of PP. Paper labels have been replaced by plastic labels, as well. This has caused some problems because the HDPE stream is now contaminated with PP, mostly from caps. On the other hand, getting aluminum out of the system has increased the usability of the recycled PET. When base cups are completely phased out, HDPE will no longer be a factor, but until then, difficulty in finding markets for the HDPE/PP blend will probably grow.

When recyclers wished to expand collection to bottles other than HDPE milk bottles and PET beverage bottles, they realized they had to instruct consumers as to which containers to put in the recycling bins, and which to throw in the garbage. If all HDPE bottles were acceptable for recycling, for example, there was no easy way to tell consumers which containers were HDPE and which were made from other resins. Recyclers needed a marking system to identify the various plastic types. As a result, the Society of the Plastics Industry (SPI) developed a coding system. It consists of a triangle formed by chasing arrows, with a number inside and a letter code underneath (see Fig. 9.1) for each of the six major packaging plastics. The number 7 stands for "other."

This coding system is now required by most U.S. states on bottles 0.47 l (16 oz.) in capacity or larger, and other containers 0.24 l (8 oz.) in capacity up to 18.9 l (5 gal.) In some states, there is no upper limit on capacity, so even 208 l (55 gal.) drums must be coded. Soon after it was introduced, the SPI codes began appearing on a variety of other types of plastic packaging, including lids, blisters, and bags. The system has received considerable criticism. The main argument is that consumers interpret the code as an indication that the item is recyclable, while in reality no recycling opportunities are available for many of the marked items. It has also been criticized as offering too little information for effective separation; for example there is no distinction between high melt flow, injection molded HDPE containers and

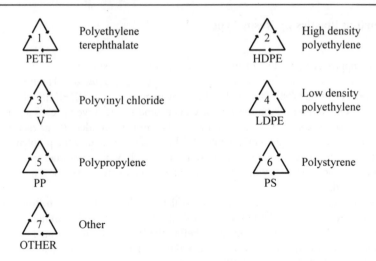

Figure 9.5   SPI coding system for plastic containers

low melt flow, blow molded HDPE containers. Efforts to reach consensus on modifications to the system failed, however. Internationally, the coding systems being developed by the International Standards Organization (ISO) do provide the more specific information needed, as well as identify a wider variety of plastics. At present, however, for most U.S. markets, the SPI codes are required.

Another approach to solving the problem of plastic resin sorting is to collect a wide variety of containers and sort them in a centralized facility. Even those recycling programs which collect only two types of plastic have to sort them. The most common approach is still hand-sorting as the material passes down a conveyor line. However, automated sorting machines have been developed which identify plastic resins and trigger an air blast to direct the containers to the desired location. Some of these systems can separate by color in addition to resin type. Systems used include x-ray, infrared, and optical scanners. They are generally limited to sorting whole or nearly whole containers, and do not work effectively on chips.

In contrast to the macrosorting techniques described above, microsorting techniques are designed to be used on chipped or granulated commingled plastics. Density-based separation of HDPE and PET is one such system. Liquids of other densities can be used to perform other separations. A system with exciting potential, although uncertain economics, is the use of supercritical fluids, where density can be adjusted very precisely to perform the desired separation. It has been reported that these systems can be sensitive enough to separate different colors based on only the small density differences induced by the coloring agents. Other microsorting techniques include froth flotation, grinding and sieving, and electrostatic processes. Differences in the softening temperatures of polymers have been used as well, where some soften and cling to a belt while others are removed. Color sorting techniques have been used to remove caps from natural HDPE, but have not been economical for more complex mixtures. An x-ray system has been developed for removing PVC flakes from other postconsumer chipped plastics [16].

For packages made from multiple resins, such as coextruded bottles, even microsorting is not usually sufficient. Molecular sorting techniques are designed to break the structures down to the molecular level by dissolution. Systems generally involve either one solvent at several temperatures, or several solvents to selectively evaporate and reprecipitate the polymers. Costs and solvent retention issues are concerns. One notable exception to the general difficulty of recycling multilayer containers into high quality materials is the PET/EVOH containers developed for catsup and other oxygen sensitive foods. These containers are produced by coinjection stretch blow molding, with two layers of EVOH sandwiched between three layers of PET. Since there is no tie (adhesive) layer, there is limited attraction between the PET and EVOH. When the containers are ground and washed, most of the EVOH is removed, leaving a relatively pure PET stream.

9.6   Degradable Plastics

In the late 1970s, there was a flurry of interest in degradable plastics as providing a more environmentally friendly alternative to nondegradable materials. As understanding of the behavior of landfills grew, it became evident that the degradable plastics broke down extremely slowly in such anaerobic environments, providing few benefits. However, in applications where composting is the disposal method, degradable plastics do have real advantages. They can also be advantageous when packaging is a litter problem.

Two types of degradability are of interest: photodegradability and biodegradability. Photodegradable plastics break down upon exposure to ultraviolet light. Thus, when left outdoors and exposed to the sun, these plastics undergo chemical changes that render them brittle. They eventually fragment into unrecognizable residues, which may or may not be biodegradable. Biodegradable plastics can be attacked and consumed by microorganisms, totally breaking down the chemical identity of the polymer. In addition, the few plastics which are water soluble are usually biodegradable once dissolved.

The first generation of degradable plastics, when they were being touted as a solid waste solution, were photodegradable. These obviously have little utility in a landfill, where there is no significant exposure to light. Next came blends of LDPE and about 7% starch, which were reported to be biodegradable. Even if these claims were true, they were of little benefit in a landfill. These blended materials have largely disappeared, as truly biodegradable plastics have emerged.

Several companies produce starch-plastic blends containing 40% or more starch, with claims that the remaining polymer is biodegradable once the starch has broken down. The degradability of starch-based thermoplastics that are close to 100% starch is well established. Many of these materials are water soluble, as well.

PVOH is a water-soluble polymer that readily biodegrades once dissolved. It has been available for a number of years, with limited applications because of its solubility and high cost. It is not melt-processible in its pure form, as it degrades before reaching its melt temperature, so it must be cast from a water solution. However, modified grades are available which can be melted and used for injection and blow molding as well as cast and blown film extrusion. Modifications to this polymer also help control its solubility, allowing the production of films soluble in hot water but not in cold, for example. PVOH is reported to be completely degraded within six months of exposure to moisture and soil bacteria [17].

Polyhydroxybutyrate-valerate (PHBV) has been produced by harvesting it from bacteria grown under carefully controlled conditions. This copolyester has been used for biodegradable films, bottles, tubes, and as coating on paper and paperboard. Degradation time for bottles in a composting environment is reported to be about 15 weeks [18].

Lactic-acid based polymers are also biodegradable. Applications include films, containers, and even closures. The material can be blow molded, injection molded, or thermoformed, as well as used as a coating for paper.

Polycaprolactone (PCL) is a synthetic, biodegradable polymer which has been available since 1975, but with only limited use. It is reported to be 95% degraded after 12 months of soil burial [17]. Blends of PCL and starch, PCL and PHBV, and PCL and nylon 6 have also been produced. Other biodegradable polyesters are also manufactured, as are biodegradable plastics based on protein, urea, and polysaccharides.

Applications where the use of biodegradable polymers makes environmental sense include fast food operations where waste is handled by composting; bags for leaf composting operations; and packaging likely to reach sewage treatment facilities or bodies of water. One interesting application is the use of starch-based thermoplastics as replacements for loosefill expanded PS. The PS "peanuts" accumulate static charges which contribute to their litter potential. Starch peanuts do not have the static cling, and also dissolve when exposed to water. While some starch-based cushioning materials were associated with an increased incidence of insect and rodent infiltration in warehouses, the purified starches used in some current materials reportedly do not attract vermin [19].

9.7 Pollutants

The production, use, and disposal of plastic resins is associated with various types of emissions into air and water. Some of the pollutants drawing the most attention to plastics in recent years are chlorofluorocarbons, dioxin, styrene, and hormone-mimicking organic compounds.

9.7.1 Chlorofluorocarbons (CFCs)

Chlorofluorocarbons were used as blowing agents in the production of some types of packaging foams, including polyurethanes and some expanded PS. After concern arose in the late 1980s about the role of CFCs in destroying ozone in the stratosphere, these materials were phased out as blowing agents in foams in the U.S. and other countries. Substitutes included hydrochlorofluorocarbons (HCFCs), which have much less ozone destruction potential than CFCs, and alternatives such as hydrocarbons and carbon dioxide, which are not ozone depleters. The use of HCFCs, by international agreement, is to be phased out worldwide in the early part of the next century. U.S. regulations already prohibit the use of HCFCs in packaging foams.

9.7.2 Dioxin

When people talk about dioxin as a pollutant, they refer to a group of related chlorinated dioxins, the most toxic of which is 2,3,7,8-tetrachlorodibenzo-p-dioxin. These compounds arise in small quantities as byproducts of some organic chemical reactions. Polychlorinated dibenzofurans are often produced as well, and have similar toxicity concerns. These two families of compounds are often lumped together and referred to as "dioxin," with their toxicities expressed in terms of equivalents of 2,3,7,8-tetrachlorodibenzo-p-dioxin. While the toxicity of these compounds is by no means fully understood, exposure to dioxin is associated, in laboratory animals, with cancer, birth defects, reproductive difficulties, and death. Dioxin has been widely referred to as ''the most toxic man-made chemical ever produced'' based on tests on guinea pigs, though hamsters tolerate 5,000 times the dosage per kilogram of body weight as guinea pigs [20].

Dioxin first drew public attention as a contaminant in Agent Orange, a herbicide used by the U.S. in Vietnam, and later was found as a byproduct of production of hexachlorophene. An incident in which dioxin-contaminated oil was sprayed in Times Beach, Missouri, leading to the death of some horses, further alarmed the public. As techniques for dioxin analysis were refined, it became evident that tiny quantities of dioxins are produced as a byproduct of various natural and synthetic processes, including combustion of organic materials. It is this fact which led to major packaging-related concerns.

When municipal solid waste is combusted, some dioxins can be produced. Dioxins are chlorinated species, so the presence of chlorinated materials in the waste is necessary for their production. PVC is a chlorinated plastic, and concerns were raised that the presence of PVC in the incinerator stream would lead to significantly greater production of dioxin. Despite a number of studies showing that control over incinerator conditions, including combustion temperature and temperatures within the emission control systems, was much more significant for the production of dioxins than presence or absence of PVC; studies showing no significant statistical correlation between amount of PVC in the waste and dioxin production; and improvements in incinerator operations greatly reducing the emission of dioxins, this has remained a public concern. Some environmentalists have called for bans on PVC production and use, related partly to concerns about dioxin [21, 22].

9.7.3   Styrene Emissions

Styrene emissions can arise from production of styrene and PS, and from residual styrene in PS packaging. Concern has arisen in the last few years about these emissions, following reports of styrene found in human and animal tissues. However, there are natural sources of styrene, and it is not clear whether production of PS is a significant factor in the body burden of this material, or whether natural sources predominate. Further, it is not clear whether there are any health consequences associated with styrene presence at the levels found. Considerable more study is needed before conclusions can be drawn about the significance of this finding.

9.7.4   Hormone-Mimicking Organic Chemicals

Relatively recently, a concern has arisen about the ability of some organic chemicals to act in the body much like natural hormones, especially estrogen. Reports of changed reproductive behavior in wildlife, feminization of male animals, and deformities in offspring have been linked to such chemicals. While most of the suspect chemicals are not associated with packaging, some, such as phthalate plasticizers sometimes used in PVC, are. Further, many of these compounds are chlorinated, adding to the current concern about the effects of chlorinated compounds, such as PVC, on the environment. Here again, the evidence is by no means conclusive. There is considerably uncertainty about the causes and significance of the observations that have been made, and much more research is needed [23].

9.7.5   Other Pollution Concerns

Plastic production is also associated with a host of more "traditional" environmental emissions, different in kind and amount, though not necessarily greater or lesser in impact, than those associated with production of other types of packaging materials. A study by the Tellus Institute published in 1992 [24] detailed publicly available information about the amounts and types of environmental emissions associated with production of several types of plastics, as well as other packaging materials (Table 9.2). These emissions were converted to dollar costs

Table 9.2   Some Emissions from the Production of Packaging Polymers [24]

HDPE	LDPE
benzene	acenaphthalene
benzo[a]anthracene	benzene
phenol	dimethyl phthalate
toluene	di-n-butyl phthalate
aluminum	ethylene dichloride
titanium	phenol
	toluene
PP	chromium
benzene	copper
bis(2-ethylhexyl) phthalate	lead
chloroform	mercury
ethylbenzene	
1,1,1-trochloroethane	**PS**
toluene	aldrin
chromium	benzene
zinc	bis(2-ethylhexyl) phthalate
	butyl benzyl phthalate
PVC	1,2-dichloropropane
acrolein	dimethyl phthalate
bix(2-ethylhexyl phthalate)	ethylbenzene
1,3 butadiene	ethylene dichloride
chloroprene	lindane
ethylene dichloride	phenol
hydrogen chloride	styrene
pentachlorophenol	toluene
phenol	chromium
propylene	lead
1,1,1-trochloroethane	
vinyl chloride	**PET**
vinylidene chloride	acrylonitrile
lead	ethylbenzene
zinc	naphthalene
	phenol
	antimony

and "scores" reported for various packaging materials (Table 9.3). As can be seen, PVC compared very unfavorably with other materials.

The Tellus study was criticized for using information which did not reflect recent changes in manufacturing processes, as well as for the methodology used to convert emissions into dollars. But this study remains one of the few publicly available reports that systematically examines environmental emissions associated with packaging. It is a prime example of the growing interest in lifecycle assessment as a guideline for making environmentally responsible packaging decisions.

Table 9.3   Tellus Institute Environmental Costs of Packaging Materials [24]

Material	Cost of material production and disposal (U.S. $)	
	per tonne	(per ton)
Paper		
Bleached Kraft paperboard	$ 487	($ 443)
Unbleached Kraft	$ 429	($ 390)
Folding boxboard from		
waste paper	$ 272	($ 247)
Glass		
Virgin	$ 173	($ 157)
Recycled	$ 140	($ 127)
Aluminum		
Virgin	$2,159	($1,963)
Recycled	$ 376	($ 342)
Steel		
Virgin	$ 403	($ 366)
Recycled	$ 394	($ 358)
Plastics		
HDPE	$ 591	($ 537)
LDPE	$ 638	($ 580)
PP	$ 662	($ 602)
PS	$ 682	($ 620)
PVC	$5,817	($5,288)
PET	$1,219	($1,108)

9.8 Lifecycle Assessment

If one desires to use the most environmentally responsible package for a product, a number of factors must be considered. One obvious part of the decision is the impacts associated with the production of a packaging material, including emissions and energy use. Impacts associated with the transportation of the material and its formation into packages must also be included. So must line operations involved in filling and sealing the package, transportation of the packaged product, waste associated with failure of the package (breakage, leakage, product oxidation, etc.). The ultimate disposal of the package should also be considered. Thus, it is necessary to consider the whole system to avoid making decisions that may be environmentally sound when the material is examined alone, but that cause increased impacts somewhere else in the system.

Lifecycle assessment is a technique for doing this evaluation. In its fully developed form, it contains at least three major components: the lifecycle inventory, impact analysis, and improvement analysis. Some authorities add another component, the scoping of the assessment to determine its goals and proper boundaries.

9.8.1 Lifecycle Inventory

The lifecycle inventory consists of a simple tally of the inputs and outputs from a product or process lifecycle. Entries are, for the most part, quantitative, but may include qualitative items (e.g., noise, habitat destruction) as well. To the extent possible, outputs are listed individually, rather than in aggregate form. Thus, measures such as masses of individual air pollutants are preferred to a measure of the total mass of air pollutants. Emissions include representative amounts of fugitive emissions as well as those occuring during "normal" process operations. The boundaries drawn around the system determine the processes included in the system, which can raise some interesting questions. For instance, should the manufacture of the tires on the trucks used to transport the materials be included?

As can be seen from Table 9.2, whenever tallies are made of the emissions associated with the production of a packaging material, different chemicals in different amounts are emitted by various alternative materials. The inventory alone does not assign relative values to these different types of emissions. It can, nonetheless, serve as a guideline in looking for process improvements, since any change which reduces emissions of a given type without increasing emissions elsewhere in the system is clearly positive. On a more qualitative basis, the inventory can identify parts of the process that are particularly problematic, in that they result in a large amount of emissions, and, with some guidance from common sense, suggest worthwhile changes even if they do result in some increase in other types of emissions. For example, a significant decrease in emissions of a toxic substance which results in relatively modest increases in a relatively benign emission is in all probability a positive change. More extensive evaluation of the impact of product or process changes, however, requires complete impact analysis.

There are some generally accepted guidelines for inventory analysis. Inventories should be complete, based on accurate information, and information should not be aggregated unnecessarily. Further, at least when used for public policy purposes, the assumptions and boundaries should be clearly laid out, and inventories should be subject to peer review and based on publicly available data [25].

9.8.2 Impact Analysis

The purpose of impact analysis is to take the inventory of product or process inputs and outputs and assign some value to it to enable comparisons between different lists and quantities of inputs and outputs. Impact analysis asks what is the effect of these inputs and outputs on the quality of the environment. Thus, impact analysis is directly connected to our value systems: what aspects of environmental quality are worth how much? Impact analysis is still very much a developing area of lifecycle assessment.

9.8.3 Improvement Analysis

Improvement analysis is used to suggest product or process modifications that can result in lessening the environmental impact of the product or process. As discussed earlier, improvement analysis, although it forms the third leg of total lifecycle assessment, need not necessarily be preceded by a full-fledged impact analysis. The inventory alone can sometimes provide clear indications of where process improvements can be made. However, when major changes which result in very different sets of inputs and outputs are contemplated, the impact analysis is needed to truly compare the alternatives.

9.8.4 Status of Lifecycle Assessment

Lifecycle assessment is still a developing technique. The U.S. EPA is carrying out research projects to further develop the usability of lifecycle assessment. Computer models for carrying out various parts of lifecycle assessment have been developed, especially in Europe. Lifecycle assessment is promulgated in some countries as a government-recognized tool for the evaluation of packaging alternatives [26].

9.9 Environmental Legislation

Legislation relating to packaging and the environment can be grouped into two general categories. The first covers legislation that impacts packaging, but is not specifically directed towards it. In this category fall laws dealing with occupational health and safety, Clean Air Act provisions, and regulations on hazardous waste disposal. These laws are outside the scope of our discussion in this book. Instead, this book focuses on the new types of legislation that are specifically directed at packaging and usually arose from concerns related to the disposal of municipal solid waste. A full discussion and description of relevant legislation is also outside the scope of this book; furthermore, these laws change rapidly. Some useful sources of information are publications by the Thompson Publishing Group [5] and by Raymond Communications [27, 28], which are updated monthly or bimonthly. All three cited contain some international information, although they are primarily directed at the U.S. market.

One complication in examining legislation dealing with packaging and the environment is that it appears at various levels of government. In Europe, there is an EU Packaging Directive, but laws differ significantly in individual countries. In Canada, there is a National Packaging Directive, but provinces have their own laws. In the U.S., state legislation is generally the most significant. In this section, these laws are grouped into general categories and their goals and status discussed, without attempting to be all-inclusive. It should be recognized at the outset that these categories are arbitrary, and many laws overlap several categories.

9.9.1 Packaging Reduction Laws

The goal of packaging reduction laws is to reduce the amount of packaging going to disposal. In general, these laws allow the target reduction to be met in a variety of ways, including source reduction, reuse, and recycling. A typical example is Canada's national policy to reduce packaging waste by 20% from 1988 levels by the end of 1992 (a goal which was met), by 35% by the end of 1996, and by 50% by the end of 2000. Other provisions are also included. As of March, 1997, reports on whether Canada reached its 1996 goal had not been issued. Predictions were that it would be very close.

A special category is reducing not the amount of packaging, but its toxicity, in an effort to reduce the environmental impact of its disposal. An example is the CONEG Model Toxics Legislation, discussed in Chapter 7.

These laws often overlap those focused mostly on mandatory recycling as a way to reduce the quantity of packaging going to disposal, and requirements for recycled content to promote markets for recycled materials.

9.9.2 Mandatory Recycling

Laws focused on mandatory recycling take a variety of forms. Some require government bodies to establish recycling programs for their citizens. Some require citizens to participate in diverting certain targeted goods from the waste stream. Others, known as producer responsibility laws, focus on the obligation of the producer of the packaging to ensure that it is recycled. The goal is to divert packaging materials from the waste stream by enabling them to be recycled.

Examples of the first type are state laws in Arizona, Arkansas, and other states which require local governments to establish recycling programs. Weaker forms of this law simply require local governments to write recycling plans, or to consider recycling in formulating their solid waste plans. A variant is laws which mandate waste reduction goals, without explicit direction as to how to achieve these goals, or in weaker form, set non-mandatory waste reduction targets.

New Jersey has an example of the second type of law, which requires local governments to establish mandatory source-separation recycling programs for at least three materials. It is a violation of the municipal ordinance to improperly place the targeted items in the waste stream instead of the recycling stream. Possible penalties include fines and even incarceration However, the usual policy for dealing with violators is to issue a series of warnings followed by refusal to pick up garbage from the offending household.

An interesting twist are laws which do not necessarily require recycling, but prohibit disposal of targeted recyclables. Wisconsin, for example, prohibits landfilling or incinerating foam PS packaging and plastic containers, among others.

Germany has the most well-known example of the producer responsibility category of legislation. The philosophy is that the packagers of the products are responsible for the ultimate recycling of the packages, and it is up to them to carry out this task. The German law sets minimum percentages for the recovery of packages and for recycling the recovered material. It also gives consumers the right to return packages to retailers, who can in turn return them to distributors, etc. The response of industry was to set up the Duales System Deutschland (DSD), more popularly known as the Green Dot system, to handle the recovery and recycling of the packages in a coordinated fashion, and avoid potentially huge messes at the retail stores. The presence of a green dot on a package signifies that the packager has paid to cover the cost of recovering and recycling the package. The consumer then returns the package to a conveniently located collection point, rather than to the retailer. The system has been widely criticized as costly and acting to restrain trade, and has had some serious financial problems. However, it is still in place and it is not likely to disappear in the near future. On the contrary, the idea has been extended in Germany beyond packaging to such consumer items as automobiles. For packaging materials, similar systems have been or are being established in many other European countries, such as France and the United Kingdom. The major difference between most of these systems and that of Germany is that they permit incineration with energy recovery to count as recycling.

The European Union Directive on Packaging and Packaging Waste is aimed at reducing the impact of packaging waste on the environment by a combination of source reduction, reuse, and recycling. It establishes specific targets for recovery and recycling, with recycling defined to include energy recovery and composting. How to reach these requirements is left to the individual countries, and most are still in the process of establishing systems and requirements. Since regulations are likely to apply to imported goods as well as those produced in the EU, the impacts will be felt around the world [29].

9.9.3 Recycled Content Mandates

The focus of recycled content mandates is to promote recycling by improving markets for recycled materials. One way to do this is to require packages to contain recycled materials. Typically, the legislation provides several options, including source reduction and achievement of target recycling rates for the containers, in addition to the recycled content provision. The targeted recycling rates are sometimes set up to increase with time, earning this legislation the designation of "rates and dates" laws. A typical example is the Oregon law which requires that rigid plastic containers either be recycled at a 25% rate, be reused or refilled, be source-reduced by 10%, or contain 25% postconsumer recycled material. The recycling rate requirement can be met either by the aggregate rate in the state, the rate for a specific type of container, or the rate for a specific product-associated container. There are exemptions for medical packaging, export packaging, and some others.

After the original law went into effect in 1995, it was amended to exempt food packaging, but not beverage packaging. When the law was passed, industry put considerable money and effort into expanding plastic container recycling in Oregon, with the result that the Oregon Department of Environmental Quality projected that the recycling rate for all plastic containers would exceed 25%, thus freeing all containers from the requirement for recycled content. A similar law was passed in California at about the same time. The state failed to agree on an official 1996 state recycling rate for plastic containers, leaving manufacturers unsure about their requirements. In California, too, the law was amended to exempt food containers. Wisconsin requires that plastic containers consist of at least 10% recycled content, with exemptions for foods, beverages, drugs, and cosmetics. There is no provision for exemption by recycling rate or source reduction.

9.9.4 Packaging Bans and Degradability Requirements

Laws which explicitly or implicitly prohibit certain types of packaging have been less popular in the last few years than they were in the late 1980s. At that time, a number of laws were passed that prohibited various types of non-degradable packaging, packaging made with CFCs, expanded PS food packaging, and plastic soft drink cans, among others. Most of those laws have since been rescinded. A major exception is the prohibition in the U.S. of the use of non-degradable plastic ring connector devices for beverage cans. This legislation addresses a litter issue rather than a solid waste issue. Animals and birds sometimes became trapped in the plastic rings when they were discarded improperly. Maine banned aseptic juice boxes, which contain combinations of plastic, paper, and aluminum foil, for a few years because it was thought that these packages could not be recycled.

Variations of these laws give government bodies the authority to consider and take action concerning packaging deemed to present a problem. Wisconsin, for example, can examine "new" packaging if complaints are received that there are not adequate markets to make recycling that type of packaging economically feasible, and can negotiate with the manufacturer to ensure adequate markets. Iowa also provides for packaging reviews.

9.9.5   Packaging Taxes and Other Economic Penalties and Incentives

Another approach provides economic incentives and disincentives for various types of packaging related to their recyclability; the institution of recycling programs; the building of recycling facilities; or the modification of processes to incorporate recycled materials. These regulations take many forms. Florida had an "advance disposal fee" levied on packages which did not meet a target recycling rate, until the law expired and was not renewed. Several U.S. states have instituted grant and loan programs to promote recycling in various ways. Michigan, for example, has provided grants and loans for community recycling programs; educational materials for children to promote recycling; equipment to establish recycling operations; and other related activities.

9.9.6   Labeling Requirements and Regulations

Some legislation relates to the information appearing on the packaging relevant to environmental considerations. One example is the plastic resin coding requirement discussed in Section 9.5.4. Table 9.4 lists the U.S. states which require the SPI code on plastic containers.

Another common type of legislation restricts environmental claims that can be made on packages. In the U.S., many of these laws grew out of the *Green Report II* [5], which was drafted after a number of State Attorneys General became concerned about what they saw as advertising misleading to consumers. In fact, a number of companies were successfully sued over advertising claims. Later, the U.S. Federal Trade Commission (FTC) also promulgated voluntary guidelines [5]. Some states have modeled state laws after these guidelines. Even when no specific environmental labeling laws exist, states can use more general "truth in

Table 9.4   States with Plastic Resin Coding Requirements for Rigid Containers [5]

Alaska	Arizona	Arkansas	California
Colorado	Connecticut	Delaware	Florida
Georgia	Hawaii	Illinois	Indiana
Iowa	Kansas	Kentucky	Louisiana
Maine	Maryland	Massachusetts	Michigan
Minnesota	Mississippi	Missouri	Nebraska
Nevada	New Jersey	North Carolina	North Dakota
Ohio	Oklahoma	Oregon	Rhode Island
South Carolina	South Dakota	Tennessee	Texas
Virginia	Washington	Wisconsin	

advertising" laws to combat misleading environmental claims. Thus, it is very important for any packager contemplating environmental claims on a package to make sure the claims are specific, accurate, and can be substantiated. Guidelines include avoiding general terms such as "contains recycled content" and using instead specific terms such as "contains at least 30% post-consumer recycled plastic."

Related to environmental labeling are programs that establish environmental emblems used as marketing tools by companies for products meeting certain environmental standards. One of the earliest of these was Germany's Blue Angel program. Japan's program is known as the EcoMark, and Canada has the Environmental Choice label. In the U.S., two competing programs, Green Cross and Green Seal, exist. Other countries also have similar programs. In general, industry participation is purely voluntary, and in fact companies have to pay a fee to be considered, gaining the right to use the label if they qualify. The types and numbers of products which can be labeled vary.

9.10 Looking to the Future

It is impossible to predict future changes in packaging and its relationship to environmental issues. Many have hoped the concern will die away, as it did for the most part after the flurry of interest in the 1970s. However, there are reasons to think this will not happen again to the same extent. First, our knowledge about the effects of human activities on the environment has increased. Second, our expectations are higher. We believe we are entitled to clean air, clean water, and protection from exposure to toxic chemicals, among others. Third, we have demonstrated our willingness and desire to do something that has environmental benefits, in spite of a little inconvenience. Fourth, there have been changes in corporate culture. Some large retail chains require suppliers to use less packaging. The automobile industry requires many of its suppliers to use reusable packaging. Corporations have adopted environmental guidelines including efforts to reduce packaging or make it more "environmentally friendly."

There are signs of decreasing immediate environmental pressures. In the U.S., the solid waste "crisis" is largely past, if it ever truly existed. Most proposed legislation to regulate packaging has failed, and some which passed has been rescinded or modified. For the last few years, there has been a trend towards less government regulation rather than more. However, recycling programs remain very popular. If these programs begin to fail because of a lack of markets, we may see renewed interest in the type of legislation which supports or mandates markets for recycled materials. This may be iminent, especially for PET, because of the drastic decline in value of recycled PET caused by the current oversupply of virgin resin.

In the final analysis, time will tell. In the meantime, the packager must remain aware of environmental considerations and factor them into packaging decisions, not in place of other considerations such as protection and market appeal, but in conjunction with them.

References

1. Boettcher, F.P. (1992) In *Emerging Technologies in Plastics Recycling*, G.D. Andrews & P.M. Subramanian (Eds.), ACS Symposium Series 513, Washington, D.C.:American Chemical Society
2. U.S. Environmental Protection Agency (1996) *Characterization of Municipal Solid Waste in The United States: 1995 Update*, EPA-530-R-96-001, Washington, D.C.:U.S. Environmental Protection Agency
3. U.S. Environmental Protection Agency (1994) *Characterization of Municipal Solid Waste in The United States: 1994 Update*, EPA-530-R-94-042, Washington, D.C.:U.S. Environmental Protection Agency
4. CONEG Source Reduction Task Force (1991) *Preferred Packaging Manual*, Washington, D.C.: CONEG Policy Research Center, Inc.
5. Thompson Publishing Group (1996) *Environmental Packaging: U.S. Guide to Green Labeling, Packaging and Recycling*, Washington, D.C.:Thompson Publishing Group
6. Woods, R. (June 13, 1995) Waste Age's Recycling Times, p. 6
7. Rabasca, L. (Aug. 22, 1995) Waste Age's Recycling Times, p. 11
8. Apotheker, S. (May 1994) Resource Recycling, p. 51
9. McCreery, P. (April 4, 1995) Waste Age's Recycling Times, p. 10
10. Person, B.D. and Schababerle, C.C. (1992) In *Plastics Recycling: Products and Processes*, R. Ehrig (Ed.), Munich:Hanser Pub., pp. 73-108
11. Greczyn, M. (Sept. 16, 1996) Plastics News, p. 11
12. The Vinyl Institute (Aug. 1993) Environmental Briefs, 4(2):1
13. Carroll, W.J. Jr., Elcik, R.G. and Goodman, D. (1992) In *Plastics Recycling: Products and Processes*, R. Ehrig (Ed.), Munich:Hanser Pub., pp. 131-150
14. Ford, T. (1995) Plastics News, p. 4
15. Bisio, A.L. and Xanthos, M. (Eds.) (1994) *How to Manage Plastics Waste: Technology and Market Opportunities*, Munich:Hanser Pub.
16. White, K. (Jan. 24, 1995) Waste Age's Recycling Times, p. 6

17. Booma, M., Selke, S.E. and Giacin, J.R. (1994) Journal of Elastomers and Plastics, *26*, pp. 104-142

18. Steigerwald, F. (1993) Paper presented at Recycle '93, Davos, Switzerland

19. Wellemeyer, J. (1996) American Excelsior Company, personal communication

20. Rawls, R.L. (1983) Chemical and Engineering News, 63(23), pp. 37-48

21. McAdams, C.L. and Aquino, J.T. (1994) Waste Age, Nov., pp. 102-106

22. King, R. (1996) Plastics News, December 2, p. 4

23. Hileman, B. (1995) C&EN, Oct. 9, p. 30

24. Tellus Institute (1992) *Tellus Packaging Study*, Boston: Tellus Institute

25. Vigon, B.W. et al (1993) *Life-Cycle Assessment: Inventory Guidelines and Principles*, EPA/600/R-92/245, Washington, D.C.: U.S. Environmental Protection Agency

26. SustainAbility, Society for the Promotion of LCA Development, and Business in the Environment (1993) *The LCA Sourcebook*, London: Business in the Environment

27. Raymond Communications (1996) *State Recycling Laws Update*, Riverdale, Md: Raymond Communications, Inc.

28. Raymond Communications (1996) *Recycling Laws International*, Riverdale, Md: Raymond Communications, Inc.

29. Fonteyne, J. (1997) In *The Wiley Encyclopedia of Packaging Technology, 2nd ed.*, A.L. Brody and K.S. Marsh (Eds.), New York: Wiley, pp. 805-808

Index